SEXY ORCHIDS MAKE LOUSY LOVERS

SEXY ORCHIDS MAKE

MARTY CRUMP

With illustrations by Alan Crump

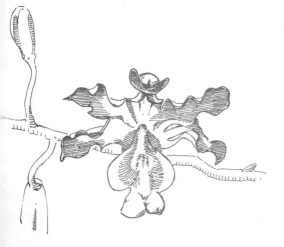

LOUSY L♥VERS

& OTHER UNUSUAL RELATIONSHIPS

The University of Chicago Press   CHICAGO & LONDON

MARTY CRUMP is an adjunct professor in the Department of Biological Sciences at Northern Arizona University. She is the author of *Headless Males Make Great Lovers* and *In Search of the Golden Frog*, both published by the Press.

The University of Chicago Press, Chicago 60637
The University of Chicago Press, Ltd., London

18 17 16 15 14 13 12 11 10 09   1 2 3 4 5

ISBN-13: 978-0-226-12185-7 (cloth)
ISBN-10: 0-226-12185-2 (cloth)

Library of Congress Cataloging-in-Publication Data

Crump, Martha L.
    Sexy orchids make lousy lovers & other unusual relationships /
Marty Crump; with illustrations by Alan Crump.
        p.   cm.
    Includes bibliographical references.
    ISBN-13: 978-0-226-12185-7 (cloth: alk. paper)
    ISBN-10: 0-226-12185-2 (cloth: alk. paper) 1. Animal behavior.
2. Animal-plant relationships. 3. Host-bacteria relationships.
4. Host-fungus relationships. I. Crump, Alan. II. Title.
    QL751.C8827   2009
    577.8—dc22

                                        2009011857

For Karen and Robert, Marty's grown children, and for Irma, Alan's wife,

with thanks for their encouragement, support, and inspiration

Contents

Preface

"GROSS!" What I had thought to be a sleeping weevil, poised on a twig in the Ecuadorian rain forest, was in fact a corpse. Its "sleepiness" was rigor mortis, and its posture expressed pure agony. Cottony white filaments smothered the weevil's contorted body. From its back sprouted a stiff, wire-like black thread, ending in a pinhead. A parasitic fungus had invaded and taken over the beetle's body. Before killing its victim, the fungus induced the weevil to move to an exposed twig, where the fungus's spores could be tossed to the wind. That night and for many nights afterward, I had vivid nightmares of suffering the same fate, smothered with cottony filaments. A grotesque, giant mushroom always sprouted from between my shoulder blades.

Nightmares aside, the interaction between insect and fungus is as natural as bluebird parents feeding their hungry chicks. Living organisms everywhere interact with members of the same and different species. Few relationships are as nightmarish as the parasitic fungus-insect interaction; many are win-win relationships breeding good fellowship on both sides. Join me in celebrating the diversity of relationships.

I've arranged this diversity into four groupings. The first involves interactions between individuals of the same species. For example, male long-tailed macaques "pay" females for sex, vampire bats share blood meals, and unborn sand sharks kill their siblings.

The second section focuses on interactions between animal species. Examples include fish that form hunting partnerships, mites that hitchhike in hummingbird nostrils, and mosquitoes that steal food from ants.

Interactions between plants and animals comprise the third section. You may be surprised at some animals that use plants for medicines, stimulants, and hallucinogens. Find out why "Mexican jumping beans" jump, and why sexy orchids make lousy lovers for wasps and flies.

The fourth section addresses interactions between the "lowly" organisms—bacteria and fungi—and plants and animals. Did you know that at least 90 trillion of "your" cells are actually bacteria? Read about the 57 varieties of bacteria (not Heinz products) that live in Komodo dragon drool. And the parasitic fungi that snatch the bodies of insects.

As in *Headless Males Make Great Lovers*, the topics I cover here are *not* scientific reviews but rather eclectic assortments of some of my favorite stories of relationships. My goal is to increase your appreciation for natural history. If occasionally I seem to suggest that non-human animals, plants, bacteria, or fungi behave with a conscious goal in mind (what scientists call teleology), or if I seem to attribute human characteristics to other animals (what scientists call anthropomorphism), please know that I am simply tripped up by my own enthusiasm in sharing these fantastic natural histories in terms to which we can all relate. Really, I know better. I most definitely do not mean to imply that non-human animals behave with conscious goals, or to attribute characteristics of ourselves to non-human animals. Readers interested in delving further into these natural history stories are encouraged to consult the references listed for each essay.

Now let's explore some unusual relationships—not as psychologists or sociologists but as naturalists. Some of these interactions occur in faraway landscapes, others in your own backyard, and still others inside your body. The relationships, from the simplest to the most complex, reinforce the reality that no individual or species exists alone. We are all interconnected.

Acknowledgments

WE ARE INDEBTED to the many scientists and naturalists who recorded the observations and experiments that document the amazing interactions described herein. Without their work, this book would not be.

Marty thanks her biologist friends and colleagues for their helpful feedback, especially her husband, Pete Feinsinger, who read every essay at least three times (and expressed great relief when she finished the manuscript) and offered constructive comments and suggestions, support, and encouragement. Steve Schuster and two anonymous reviewers read the manuscript for the University of Chicago Press and offered helpful suggestions. Marty also thanks Steve Schuster and Heather Bleakley for illuminating discussions regarding multilevel selection. Kurt Auffenberg, Martha Groom, Peter Price, and Tom Whitham provided information and/or references. Non-biologist members of Marty's writing group convinced her that not every detail was needed (or even interesting) and helped make her prose more accessible to a general audience. For this, she thanks Martha Blue, Nancy Bo Flood, Steve Hirst, Kay Jordan, and Marilyn Taylor. Robert Feinsinger, Karen Hackler, and Judy Hendrickson also offered insight on the text from a non-biologist perspective. Marty thanks her brother Alan for coming on board a second time with his creative illustrations. Finally, Marty thanks Conan, her long-haired dachshund, for providing comic relief at just the right times—and for tolerating occasional dinner delays and forgotten treats while Marty stared at the computer screen.

Alan gratefully acknowledges Marty for her generous spirit in creating opportunities for him—from adventures in the rain forests of the Amazon to the wilds of book illustration. Alan also sincerely thanks his wife, Irma, for her encouragement, support, and continuous inspiration.

Once again we thank our editor, Christie Henry, for her enthusiasm and insight throughout this project. And thanks to Erin DeWitt, superb copy editor.

1 Whatever Happened to Baby Booby? & Other Interactions among Animals of the Same Kind

CONSIDER THE RELATIONSHIPS you have with other people. You might interact with grandparents, parents, siblings, your spouse, your children, friends, neighbors, your boss, coworkers, and strangers—all in the same day. You might depend on other people for food, shelter, protection, spiritual guidance, legal or financial advice, learning, emotional support, transportation, medical care, child care, and a host of other services. Some of our interpersonal interactions are mutually beneficial. Sometimes only one individual gains. And sometimes one person gains while the other gets hurt.

Animals of other species also interact with one another in many ways. The most basic interaction is sex. Even most animals that live solitary lives get together once in a while—to mate. But mating isn't always harmonious. We'll look at what happens when males and females have different ideas about mating: "battles between the sexes" that involve resistance, coercion, and even enforced chastity. Next we'll focus on animals that sometimes put humans to shame when it comes to sustaining long-term partnerships: birds. About 90 percent of all bird species are bonded to a single partner. That doesn't mean all are sexually faithful, though. Far from it. As you'll see, long-term partnerships and adultery often go hand in hand. Some birds even "divorce" their mates. Still, others stay together and are sexually faithful their entire lives.

Sex isn't everything. Social animals often cooperate outside of mating, for example, to remove ticks and other foreign material from the skin, fur, or feathers of relatives or buddies. They may form hunting partnerships or defend each other from predators. Some animals provide child-care services: they babysit young that are not their own. Others offer food to friends and relatives. Each of these themes involves fascinating and sometimes quirky behaviors.

Not all relationships are positive. Sibling rivalry isn't confined to humans, and certain animals take it to an extreme: they kill and may even eat their brothers and sisters. Be glad you weren't a late-developing sand shark embryo, the fastest-developing Cuban treefrog tadpole in the family, or the second-hatched blue-footed booby.

NOT TONIGHT, HONEY

Nobody will ever win the battle of the sexes.
There's too much fraternizing with the enemy.

HENRY KISSINGER

We joke that from a reproductive standpoint men are "expendable" because they can contribute to reproduction often and over a long period of time. They're a dime a dozen. In contrast, females are more "valuable" because they can reproduce only a small number of times in their lives.

Is there any biological basis to the expendable/valuable argument? Yes, and humans aren't alone in this regard. Bear with me a moment before we get to the stories. One commonly held viewpoint is that sex role is determined largely by initial investment in gametes, or sex cells. Males produce lots of tiny sperm. Females produce few, energy-rich eggs; thus, their initial investment in offspring is much greater than that of males. A male can maximize his paternity potential by mating with as many females as possible. A female, though, often needs only one male per reproductive season to fertilize her eggs. Whereas a female might benefit by choosing her mate carefully, it often pays for a male to sow his wild oats widely. What this means is that males often compete for the limited number of receptive females.

In nature when males try to mate as often as possible and compete with each other for limited encounters while females hold out for the "best" males, the differing goals can lead to conflict between the sexes. Sometimes the conflict involves coercion, manipulation, deceit, and even physical harm by one sex to the other. And the other sex doesn't just take this lying down. Consider the following non-human examples.

ONE SUMMER I STUDIED aggressive behavior in variable harlequin frogs (*Atelopus varius*) in the mountains of Costa Rica. My study site was a forest stream where the frogs congregated on boulders in the water and on the ground nearby. The frogs were active during the day, and as long as I stood or sat quietly, I could watch without disturbing them. Each variable harlequin frog has a unique black-and-yellow color pattern. I took a Polaroid picture of each frog and kept a mug file so that I could recognize individuals.

Like most male frogs, a male variable harlequin climbs onto the larger female and clasps her with his forelimbs in a position called "amplexus." In most species of frogs, the pair stays in amplexus for a few to 24 hours before the female lays eggs and the male fertilizes them externally. In the variable harlequin, how-  ever, the male stays locked in amplexus for days or weeks! Because piggybacking males sit too high off the ground to capture much food, toward the end of the breeding season they become frightfully emaciated. One wonders how these weakened males can still get excited enough to fertilize their mates' eggs!

Why might males hang on so long in amplexus? On any given day, the sex ratio of frogs out and about at the stream was strongly skewed toward males. Most of the females were probably hiding in rock crevices. Perhaps once a male encounters a female during the breeding season, he mounts and hangs on because he might not get another opportunity to mate. But why should a female put up with lugging around a male for days or weeks when there are plenty of males to go around?

One afternoon as I sat on a mossy boulder near the spray zone of a two-foot waterfall, the answer became obvious: Females don't put up with it. I spotted movement on the rock face: a pair of amplexing harlequin frogs. The female slowly lifted one hind limb, rolled a bit to the opposite side, and tried to dislodge the male. For the next four hours, she worked to shed her piggybacking suitor. She crawled into a crevice and tried to scrape him off by rocking back and forth. When that didn't work, she crawled back out and bounced up and down like a bucking bronco. The male held tight. A week later I found the same pair on the same rock face. During the 30 minutes I watched them, the pair sat passively. The female might have been too exhausted to fight back, was taking a breather, or was nearly ready to lay eggs.

Over the next few weeks, I watched other females try to dislodge males. In each case where the female was successful, the dislodged male dismounted over the female's head. Big mistake. Payback time. Each time the female pounced on the male and jumped up and down, pounding his head against the ground or rock with her forelimbs.

Clearly, prolonged amplexus presents a conflict of interest for the sexes. It makes sense that a male should nab a female when he can, but if the female's eggs are not mature yet, she is stuck lugging around a deadweight for days or weeks. It's difficult to say who's winning the battle of the sexes in these harlequin frogs. From what I observed, it was a tie. About half the males hung on. The rest got dumped and stomped on.

So, what does a female gain by attempting to dump a male that has jumped on too soon? One possibility is that by resisting males, females end up with the strongest and most tenacious guys around—great genes to pass on to the kids. Alternatively, male quality might not be involved at all. Perhaps females not yet ready to lay eggs resist amorous males simply to avoid wasting energy lugging them around.

HAVE YOU EVER watched water striders, those long-legged, slender insects that skate on the surface of ponds or slow-moving streams? In many species, males use either their antennae or legs to grasp reluctant females during mating. Once a male grabs a female, he hangs on and she is stuck carrying him— just as in variable harlequin frogs. She skates for the two of them, which costs her 20 percent more energy. Also, because she now skates more slowly and is less agile than when alone, she is both more likely to get eaten and is less efficient hunting for food.

Her defense? Female water striders have antigrasping structures. For example, females of some species have elongated spines that flank their genitalia and discourage unwanted suitors. Females that try to resist grasping males might come out ahead. As with variable harlequin frogs, the females' eventual partners might be those males strong enough or persistent enough to overcome females' resistance. Or resisting females might simply live longer and/or find more food.

So, who's currently winning the arms race in water striders: persistent males or resisting females? In those species where males have exaggerated grasping structures and females have exaggerated antigrasping structures, it may be too close to call. In some species, male grasping structures are stronger than female antigrasping structures, and mating rates are high. Score one for males. In other species where the reverse is true, mating rates are low. Score one for females.

A MORE EXTREME example of sexual conflict involves bedbugs. These flat ⅕- to ¼-inch-long bugs look remarkably like apple seeds—same color and shape. Unlike most apple seeds, though, human bedbugs make disagreeable houseguests. After dark they crawl out from crevices in bedding and mattresses and gravitate toward warmth and carbon dioxide: sleeping people. They pierce skin and suck blood.

Regardless of how we might feel about the feeding habits of bedbugs, their reproductive habits are fascinating. Males of species with internal fertilization normally insert their reproductive organs into the females' reproductive tracts during copulation. Not so with bedbugs. They display "traumatic insemination."

Sounds nasty, doesn't it? It is, from the female's perspective. A male bedbug mounts the female sideways, grasps her with his legs, curves his abdomen under hers, pierces his dagger-like external genitalia through the underside of her abdominal wall, and ejaculates sperm and fluids into her body cavity. Sperm travel from the female's blood into storage structures, then on to the ovaries, where the eggs are fertilized. Although the female's reproductive tract is fully functional, it ends up being used only for laying eggs. You're probably wondering why. This unusual mating system may have evolved as a way for males to overcome resistant females. It seems that male bedbugs are way ahead of male harlequin frogs and water striders—if you want to look at it that way.

What about the female bedbug's point of view, though? Can traumatic insemination hurt her? Yes. She might experience blood loss, infection, or an immune reaction to the sperm and fluids introduced into her blood. In addition, wound repair and healing require energy that could be spent on something else, such as foraging for more blood. Females forced to mate repeatedly don't live as long as less molested females. Is there anything female bedbugs can do to resist males? Like female variable harlequin frogs, females of some kinds of bedbugs vigorously shake males that try to mount them. If successful, the females run away.

Even more remarkable is that females of many advanced species of bedbugs have a secondary reproductive system called the "paragenital system," which consists of one or both of two parts. The ectospermalege is a region of swollen and often folded tissue centered in the abdominal wall where males would normally try to pierce females. This tissue provides the female with some protection from the stab. The mesospermalege, located underneath the ectospermalege, is a pocket or sac attached to the inner surface of the abdominal wall. This sac receives the ejaculate if the male succeeds in penetrating. As a point of interest, the human bedbug has both structures. Experiments suggest that these structures reduce the direct costs of piercing trauma and infection by pathogens introduced with the piercing. In an evolutionary sense, females have fought back.

LET'S GO TO some less abusive relationships. Females of some animal species having internal fertilization mate with more than one male to fertilize a cycle of eggs, but some of those presumptive fathers may not sire any of the eggs. In some species, males have a mechanical way of ensuring that their own sperm will indeed hit the jackpot: copulatory (mating) plugs that serve to enforce chastity from then on. In rats, guinea pigs, squirrels, and other rodents, this plug is formed in the female's vagina by a coagulating substance in the male's seminal fluid. Males of garter snakes and some other snake species produce copulatory plugs composed of proteins and lipids from their kidneys. They insert these plugs into the females' cloacae following insemination. Certain insects have ingenious copulatory plugs: their own body parts. After mating, a female biting midge eats her mate, but his genitalia stay lodged in her genital opening and provide a plug that inhibits other males from inseminating her. When a male honeybee catches a virgin queen during her nuptial flight, he too gives his all. His genitalia explode inside the female. Leaving his privates inside to block the queen's vagina, he falls to the ground and dies.

A copulatory plug might assure paternity for the male, but what good is it for the female? It might prevent her from being hassled by other males, but it might not always be a benefit. What if she were later to come across a "better" mate? Or what if mating with additional males might increase the genetic diversity of her offspring? In species with female promiscuity, there's often a conflict of interest between the sexes: it's a male's advantage

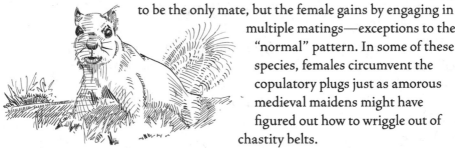

to be the only mate, but the female gains by engaging in multiple matings—exceptions to the "normal" pattern. In some of these species, females circumvent the copulatory plugs just as amorous medieval maidens might have figured out how to wriggle out of chastity belts.

John Koprowski spied on the sex lives of fox squirrels and eastern gray squirrels on the campus of the University of Kansas. He observed that following copulation, females groomed their genitalia. While cleaning themselves, they often removed copulatory plugs with their incisors. Sometimes the squirrels ate their plugs; sometimes they threw them onto the ground. By removing the plugs, females could later mate with additional males.

It's to a female honeybee's advantage to mate multiple times, to store more sperm. But what's the use of re-mating if she is plugged up with a previous lover's genitalia? Male honeybees have evolved a way of dislodging their predecessors' copulatory plugs. This allows a "Johnny-come-lately" male a stab at paternity, but because it also increases the quantity of sperm the female receives, both sexes benefit. Think about it, though. The "Johnny-come-lately" male gains by not having his exploded genitalia removed in turn by a successor, whereas the female gains by having it removed so she can accumulate more sperm. There's conflict between the sexes again.

So, how does a male honeybee remove another male's genitalia? If you ever have occasion to examine a male honeybee's phallus, look for the hairy structure at the tip. That's what he uses to try to gouge out the previous male's privates—"try," because not all attempts are successful. After all, if genitalia-gouging were 100 percent successful, it wouldn't ever pay for our male to leave behind his own privates unless he could be sure that he was the "Johnny-come-latest!"

THE BATTLE OF the sexes is a dynamic evolutionary process. At a given point in time, it might appear that one sex is ahead in the running battle. But give the other sex time, and the odds will probably even out. Henry Kissinger was right. Neither males nor females will win the battle of the sexes. We need each other—even though at times the opposite sex might act like a different species.

Women are from Venus; men are from Mars—or so we thought in middle school. Men's and women's physiologies and anatomies are regulated by different chemicals. Our brains are wired a little differently. We often view the world—and life itself—from different perspectives. Perhaps we need each other to fill in the gaps, to feel complete. So try to understand when he forgets your anniversary. Or when she spends an hour primping before going out to dinner. Of course, there are many exceptions to these stereotypes. To begin with, not all males are promiscuous and not all females are coy. Once we accept the basic differences between men and women, though, we find beauty and mystery in our relationships. Maybe that's why we keep fraternizing with the enemy. In humorist Dave Barry's words:

What Women Want: To be loved, to be listened to, to be desired, to be respected, to be needed, to be trusted, and sometimes, just to be held. What Men Want: Tickets for the World Series.

TO HAVE AND TO HOLD

To have and to hold from this day forward,
for better for worse,
for richer for poorer,
in sickness and in health,
to love and to cherish,
till death us do part.

Modernized from the *Book of Common Prayer* (1549)

The traditional marriage vow is a promise "to have and to hold" only one person forever. This promise sometimes endures; other times it's broken. Likewise, some non-human animals that form long-term partnerships are faithful (as far as we can tell) until "death us do part." Others are not.

An idealistic college student approached me in the late 1970s after my lecture on monogamy in birds. She gushed that she admired birds for their faithfulness to their partners and regretted that humans often could not do likewise. I had just explained that monogamous birds typically had exclusive mating relationships. Now we know differently. Thanks to recently developed molecular techniques that allow us to determine parentage, we've learned that many animals form partnerships without being sexually faithful to each other. If I could locate that long-ago former student, I'd point out that even though some birds do form faithful long-term partnerships, others commit adultery and even "divorce" their mates.

Many biologists now use the term "monogamy" only in the social sense of a prolonged association between one male and one female. How long is a "prolonged association"? The length of a monogamous relationship depends on the species. In some it's the time it takes to raise one brood; in others it's a lifetime. Other biologists argue that "social monogamy" is oxymoronic and that genetic monogamy (where two partners have offspring only with one another) is the only way in which the term "monogamy" should be used to classify mating systems. To avoid this argument, I will simply use "persistent pairs" to refer to social situations in which males and females remain together whether or not they remain sexually faithful to each other.

ABOUT 90 PERCENT of the 9,000 or so bird species form persistent pairs. Pair bonds increase cooperation between male and female and enhance the pair's ability to survive and breed successfully. Both individuals may incubate the eggs, feed the chicks, and watch for predators. Some birds form continuous partnerships that last throughout the year. Other partnerships last only during the breeding season.

Barnacle geese form quite impressive continuous partnerships. Once paired, the male and female often stay together day in and day out for their entire lives. Most barnacle geese pair for the first time at one or two years of age. They choose their mates carefully from thousands of possibilities. Like humans dating, these geese enter trial relationships, sampling up to six potential mates before settling on a permanent partner. They hang  out together for up to several weeks and "decide" whether they can make it as a team, staking out space and finding food on the wintering grounds. If the relationship doesn't work out, the couple breaks up and returns to "dating." Barnacle geese prefer to pair with familiar individuals born in the same area, the same year. Why? Such pairing might make life easier. Biologists speculate that familiar geese might better coordinate their future breeding and feeding activities, simplify their courtship displays, bond more readily, and live together with less conflict than individuals that begin a relationship as complete strangers.

Jeffrey Black and his colleagues studied a population of thousands of barnacle geese from 1973 to 1991. These birds bred mostly on the small Norwegian islands of Svalbard during the short arctic summer and then flew south to winter on coastal marshes and grasslands along the coast of Scotland and northern England. The investigators found that 99.6 percent of the geese formed persistent pairs. Sixty-five percent of 2,618 birds had only one mate during their lifetimes, which averaged 8 years. Most terminations happened because a partner died. Some barnacle geese live for 24 years; pair bonds persisted in some of these long-lifers for 16 years!

What keeps most barnacle geese together for so long—not only while breeding but also all day, every day, on the wintering grounds? There are lots of reasons. The geese need teamwork to compete for food on the wintering grounds so the females can fatten up before migrating and producing eggs. A female with a partner can spend more time foraging, thanks to the male's protection, and therefore can store more energy reserves. Once a pair returns to the arctic to breed, they work as a team to compete with other pairs for patches of food and nest sites. Males protect their mates, and both partners protect their young from hungry gulls and arctic foxes. Long-term pairs can better monopolize positions at the edge of a flock with best access to grasses and sedges. Males benefit from long-term partnerships because they have mates ready to breed once they're back in the arctic. No need to get back into the "dating" game.

Barnacle geese clearly surpass humans when it comes to durable relationships. Maybe partnering up with our nursery-school and kindergarten playmates would help us maintain long-term relationships. On second thought, maybe not.

SOME BIRDS THAT form persistent pairs copulate outside the pair bond; others don't. In large part, the difference in behavior depends on availability of additional mates, though species that need both parents to care for the young tend to be less promiscuous. Typically, both males and females are most likely to be promiscuous when receptive individuals of both sexes are readily available nearby. A female that accepts or even solicits copulations from other males might benefit in several ways. Promiscuity might offer a hedge against infertility in her social partner. It might increase the genetic diversity of her young. Theoretically, it might allow her to have young that are fathered by higher-quality males than her social partner, though there isn't much evidence for this scenario. For males, relationships "on the side" simply increase their reproductive success by allowing them to sow their wild oats more widely.

Australian splendid fairy-wrens provide a great example of partnership without the constraint of sexual fidelity. These small, insectivorous birds live in dry, shrub areas of Australia. Couples stay together for life, but they mate more often with other birds than with their own mates. "Sluttish," some of us might grumble. Others might call this "free love" liberating.

Male splendid fairy-wrens often fly into neighboring territories. These philanderers perform striking courtship displays to females, showing off their brilliant blue and black feathers, on occasion carrying pink or purple flower petals as they approach. Sometimes they score, sometimes not. How do these birds benefit from long-term pair bonds that include frequent sex with others? To answer this question, we need to look at both the birds' social interactions and their real estate.

One adult male and one adult female live together year-round in a lifetime partnership. Often their male offspring can't find places to set up their own breeding territories, and female offspring can't find eligible bachelors with territories. So the offspring become "helpers." They hang around and help defend their parents' territory and care for younger siblings. The family group, consisting of up to eight fairy-wrens (one senior male, one breeding female, and helpers), forages together and roosts together at night. But males—both patriarchs and helpers—sneak off to visit and mate with neighboring females. Promiscuity is so prevalent that a genetic analysis of 30 broods revealed that between 65 and 100 percent of the chicks could not have been fathered by any male within the group to which the female belonged!

Fairy-wrens display another unusual social arrangement: helpers often replace senior males or breeding females within their own group. One study found that 40 percent of replacements came from within the group. Thus, partners of a social pair are often brother-sister, mother-son, or father-daughter. But since most fertilizations come from copulations outside the pair, there's minimal effect of inbreeding from partnering with a close relative.

Now the real estate part of the answer. Splendid fairy-wrens live in fire-prone shrub habitats. Often all the suitable habitat is occupied by territories of different family groups. After a fire, a group may abandon its territory, but as the vegetation recovers, birds quickly fill the vacancy. When one member of a partnership disappears, the survivor stays in the territory. He or she doesn't wait long—sometimes only a few hours—before either a helper from within the group or from a neighboring territory fills the partner position. Because of limited space, the spot where a bird can find a breeding vacancy might dictate the choice of the long-term social partner. But what if the social partner isn't a compatible sexual partner? No problem. Enter promiscuity. Separate the social partner from the sexual partner, and the birds have freedom of choice.

As unromantic as these birds' relationships may seem, pair bonds in splendid fairy-wrens might persist because of scarce real estate—the permanent territory. Both male and female need the familiar area of a territory to forage efficiently. The female needs that space for raising her young. The area also provides a focal point where neighboring males can find her. The male needs the territory because that's the only way he can gain a social partner. Although he doesn't fertilize all her eggs, he still increases his reproductive success by having an older, more productive partner with whom he can sometimes mate. Besides, to philander with neighboring females, he needs a home base. Not only does partnership without sexual fidelity represent a workable system for these birds; their environment also seems to encourage it.

IN MANY BIRDS, pair bonds can potentially end at any time. The members of a pair might separate partway through the breeding season or simply not pair up again the next breeding season. Former partners that don't continue to breed together are said to have "divorced." (Many biologists use the term "divorce" for this separation in non-human animals. They do not intend to imply that birds feel the same way about splitting up that humans do—or that birds hire lawyers. To avoid the word's unfortunate connotation, other biologists use terms such as "mate switching," "pair splitting," or "nonretainment." Either way, to perform well as scientists, we try to retain emotional distance from our subjects and not interpret their behavior in either teleological or anthropomorphic terms.)

After "divorce," birds usually re-pair with new mates. Divorce rates vary widely among species that form long-term partnerships, from virtually zero percent in populations of Australian ravens and wandering albatross to almost 100 percent in house martins and greater flamingos. Divorce can come about for various reasons.

Divorce can happen because one partner deserts the other. If predators destroy their first nests, female white-tailed ptarmigans move to other territories and switch partners. A population of willow tits studied in northern Finland over a period of seven years averaged an annual 12 percent divorce rate. One-year-old females deserted their mates more often than did older females. Divorced females usually paired with older males having higher ranks in the winter flocks than did their previous partners (gold diggers!). Thus, divorce for willow tits may increase a female's reproductive success.

Sometimes outsiders separate the partners. Oystercatcher partnerships sometimes end in divorce when usurpers chase either one or both mates from their territory. A male might chase out a male, a female might oust a female, or a neighboring pair of birds might usurp the nesting territory of a pair. In the last case, the displaced pair usually separates.

Tardiness can also cause divorce. It might not pay for the first bird of a pair arriving on the breeding grounds to wait long if its mate doesn't show up on time. If the mate has died, the surviving bird would waste precious time waiting for its "lost love" and miss out on obtaining a new mate in time to breed. Better to take a new mate who's already on the breeding grounds. This sort of divorce happens in many species that migrate to and from breeding sites, including Adélie penguins and willow ptarmigans.

Theoretically, one member of the pair could simply reject its partner and chase it away, human-style, to initiate divorce. Curiously, such behavior has not been observed for birds in their natural environment.

DESPITE THESE EXAMPLES, my former student might still consider birds more admirable than most humans. After all, we promise "To have and to hold from this day forward . . . till death us do part." Yet about 38 percent of marriages in the United States end in divorce. And unlike birds, we often engage in active mate rejection, not just desertion. Perhaps lifetime partnership in non-human animals surprises and intrigues us because we ourselves seem to have such a hard time staying with one partner, even socially.

YOU SCRATCH MY BACK, I'LL SCRATCH YOURS

As a practical and satisfying way to express affection and get rid of dead skin and parasites, grooming is unbeatable.

BARBARA SMUTS, "What Are Friends For?"

Social animals help relatives and neighbors. In this and the next three essays, we'll look at four ways that members of the same species cooperate: keeping clean, hunting for food, taking care of the kids, and defending against predators. First, personal hygiene.

Most birds spend hours each day preening their feathers to remove bacteria, fungi, parasites, and dirt that render feathers less effective for insulation, waterproofing, flying, and social communication. Watch a bird

perched on a tree branch and you'll likely see it fluff its feathers and then comb them through its bill. It might squeeze waxy oil onto its bill from a gland at the base of its tail and then spread the oil over its feathers. The oil keeps the feathers from drying out and deters fungal and bacterial growth. Birds scratch with their feet to preen hard-to-reach areas on their heads. Some very social species such as babblers, waxbills, parrots, pigeons, and many nesting seabirds preen each other, a behavior called "allopreening."

Red avadavats, social finches common in parts of India, live in flocks for most of the year. When inactive, they clump together on perches and often lean against one another. They sit still, drowsy eyes partially closed, or they preen themselves. But they also "ask" others to preen them. A bird that is soliciting preening ruffles the feathers on its chin, crown, and nape; makes nibbling movements; and may give a high-pitched call. A neighbor responds by preening the bird's head. The individual being preened often rolls its head forward, backward, or to the side, presenting different areas to be serviced. The preener grasps each feather at the base and passes the shaft through its mandibles as it nibbles.

Experiments with red avadavats reveal that the amount of allopreening does not depend on how badly the head feathers are soiled. There's more involved than just hygiene. Each bird clumps with and allopreens only certain acquaintances, suggesting that allopreening helps to strengthen social relationships. The tactile stimulation provides a social reward.

Jungle babblers from India live up to their name. They shriek, cackle, gurgle, rattle, squawk, and wheeze. Jungle babblers live in groups of up to 20 individuals. Their babbling enables the birds to warn each other of predators, attract members of the same group, and interact with other groups. Some individuals stay with the same group for several years. These highly social and cooperative birds preen each other. Often one solicits the service by pecking others and then erecting its head and neck feathers. While being cleaned, the recipient usually crouches low on the branch, raises its head, and stretches its neck. The preener jabs its bill into the

feathers, removing dirt particles and external parasites. Unlike red avadavats, babblers allopreen all the way from the head to the rump. Most solici-
tation is done by lower-ranking birds toward more dominant birds, suggesting that soliciting and receiving allopreening helps to maintain a less
dominant individual's position within the group. In human terms, this
would be like a thirteen-year-old girl asking a senior cheerleader to style
her hair. The younger teen might hope to be accepted into the cheerleader's
social circle by making the older girl feel important and superior.

IN NON-HUMAN PRIMATES, mutual grooming, or allogrooming,
improves hygiene. But that's not all. Other possible benefits include social
bonding, coalition building, appeasement and reconciliation, relief from
stress and boredom, food, babysitting service, and sex.

Female yellow baboons of Amboseli National Park in Kenya spend
more time in grooming each other than in any other social activity except
for infant care. Adult females groom their young, their siblings, and each
other. One baboon approaches another
and lowers or directs her chest, neck,
or cheek toward the prospective
groomer. The "groomee" cocks her
head and avoids eye contact with the
other female—critical if the groomee is
higher in social rank than the prospective
groomer. Even a quick glance from her
is likely to be perceived as a threat, and
the prospective groomer will flee. If all
goes well, though, the groomer combs her
fingers through the other's fur, extracts ticks and other parasites, and
eats them. Etiquette dictates an exchange of roles every few minutes. Non-
reciprocating baboons and those offering substandard grooming have
trouble finding future grooming partners.

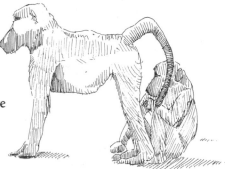

Why all the grooming? Disease-carrying ticks plague the baboons. Adult
ticks perch high on grass stems and leap onto animals that brush against
them. Baboons pick up ticks while resting, feeding, and sunning themselves
in the morning beneath their grove of sleeping trees, and while indulging in
afternoon siestas on the ground. Not surprisingly, the baboons frequently
groom during these times. After yellow baboons pass through tick-infested
habitats, females often allogroom, possibly enabling them to remove ticks
before those embed in the skin. Males generally de-tick themselves.

Let's move on to another primate. Michael Gumert discovered that male longtailed macaques from Indonesia "pay" females for sex by grooming them. In his studies, females were nearly three times more likely to mate and engage in other forms of sexual activity—including mounting, genital inspection, and presenting her hindquarters—if males groomed them first. Sexual activity occurred during or after a grooming bout, and males groomed their partners longer when sexual activity was part of the deal. Males didn't groom just any female. From a sample of 243 male-to-female grooming bouts, 89 percent were directed toward sexually active females.

Gumert interpreted the results of his 2007 study within the context of social markets—the idea that individuals trade social behavior. The theory assumes two classes of social partners: "holding" and "demanding." Holding individuals hold access to a commodity. Demanding individuals seek that commodity. This difference results in an exchange where the demanding class offers something to the holding class to gain access to that commodity. Theoretically, this system of trade should follow basic principles of economics, including supply, demand, advertisement, and value. When the holding class is scarce, the demanding class should offer more for the commodity. When the holding class is abundant, the demanding class should offer less because the commodity is less valuable.

Gumert's study is the first to provide evidence of market-like trading for sexual activity in animals. Females are the holding class, and their  commodity is sex. Males are the demanding class, seeking access to sex. Gumert found that the macaques adjusted their behavior depending on the market. Length of grooming-mating interactions depended on the number of females nearby. When fewer females were around, males spent longer grooming their social partners. The "mating fee" became cheaper when females were more available. Clearly, supply and demand affected the value of the commodity—sex.

Dominance also affected the social market. Higher-ranked macaque males not only mated more often than did lower-ranked males, but they also spent less time grooming their social partners during grooming-mating interactions. Gumert speculated that high-ranked males might pay less for sex because their social power gives them easier access to females. Females might demand less grooming from them because higher-ranked males might be more valuable partners. For example, such males might be better defenders both of their mates and any resulting offspring than are lower-ranked males.

Female longtailed macaques groom males, but seemingly for a different reason. Female-to-male grooming did not increase the likelihood of sexual activity. Gumert suggested that females seek social services other than sex when grooming males. They may groom to appease males and/or to maintain social relationships because bonded male "friends" might later protect their offspring.

WHILE WORKING IN Ecuador, I sensed the social bonding function as well as the hygienic benefit every time I saw allogrooming within human indigenous cultures. Allogrooming was often done sensuously, much the way you or I might run our fingers through a lover's hair. During a Quechua wedding ceremony, a young girl sitting on the bamboo floor of the hut searched through another young girl's hair. Each time she found a parasite, she popped it into her mouth. While riding upriver in a *piragua* (canoe), a Cofán woman groomed her son's head, then turned and did the same for her husband. Several times she flicked objects into the water, so I assume she was successful. A Chachi woman stopped her basket-weaving, reached over to the hammock a foot away, and checked her friend's head. After finding several ticks, she ripped them apart between her fingers and continued weaving. Her friend then searched through the weaver's hair. No words spoken, but the gesture was reciprocated.

Have you ever returned from a hike and done a "tick check" on your partner or child and then removed the ticks you've discovered? Have you ever deloused another person by running a fine-toothed nit comb through his or her hair to remove egg cases of head lice? Brushed dandruff, lint, or hair off a friend's clothing? We allogroom our children and incapacitated adults: brush and floss their teeth, clean their noses and fingernails, wash their hair and bodies, change their diapers. We pay, sometimes extravagantly, for professional allogroomers, including dental hygienists, dermatologists, cosmetologists, and pedicurists. Nail salons and beauty shops provide other allogrooming services. Humans groom each other with at least as much fervor as our non-human primate relatives.

BY DEFINITION, social animals cooperate with each other. Whether cleaning a relative or unrelated neighbor yields hygienic benefits or ex-

presses affection, often the expectation is that the service will be reciprocated. You scratch my back in that hard-to-reach spot, and I'll scratch
yours (and maybe eat what I find in the bargain)!

BUBBLE BLOWERS, POTHOLE PLUGS, AND OTHER GROUP HUNTING ROLES

With precise timing, two or more [bottlenose] dolphins herd a school of fish
toward a mud flat at low tide. The massive bodies of the dolphins, which often
weigh 300 to 400 pounds, push up a wave that bursts onto the beach. Fish
caught in the wave are stranded, and the rush carries the dolphins out of the
water. They roll to the right, dig in a pectoral, and . . . snap up the fish, as if they
were pieces of popcorn. . . . After their picnic on the beach, the dolphins slide
back into their element and gracefully glide away.

ALAN P. TERNES, *"Picnic à la dauphine"*

Bottlenose dolphins each eat about fifteen pounds of fish every day. They
click and whistle, echolocating to find fish and to communicate with one
another. These dolphins often trap their prey by driving them toward a
barrier such as the shore or toward other dolphins that close the trap by
encircling the fish. Two bottlenose dolphins sometimes trap fish between
themselves, swimming alongside the prey and converging on it until it has
no place to go.

Stefanie Gazda and her collaborators recently studied bottlenose
dolphins off Cedar Key, Florida. They reported that individual dolphins
consistently carry out specific tasks during successive fishing expeditions.
This behavior, called "role specification," is extremely rare in mammals.
Division of labor isn't that uncommon, but individuals usually change
roles. The investigators found that within a group of three to six dolphins,
one individual—the "driver"—consistently herded fish in a circle toward other dolphins, that created a barrier, staying tightly bunched
together. Trapped fish leapt into the air, and all the dolphins seized their
airborne prey.

Other sometime group hunters are
humpback whales that gather to feast
on herring. Often three to seven work
as a team. Most individuals are "herders."
When they find a school, the whales rush at
the fish. The herring swim upward to escape.
Meanwhile another whale, the "bubble blower,"

swims in a circle above the school continually blowing bubbles. The bubbles form an ever-narrower "net" around the herring, and the net channels the fish to the surface. At that point, the 40-foot giants, mouths agape, lunge upward through the school, fill their gullets with herring and water, and erupt from the surface. Highly coordinated, the whales lunge within a second or two of each other. They close their mouths, and their huge tongues force water out through the baleen (filter-feeding apparatus). What's left is a mouthful of fresh fish.

Humpbacks in groups no doubt feed more efficiently than do solitary individuals. And they can eat a wider variety of prey. For example, cooperative feeding allows the whales to exploit a rich source of energy— herring—that is less easily captured by solitary feeders.

Some birds also herd fish. White pelicans, majestic birds with nine-foot wing spans, eat freshwater fish such as perch, chub, and carp. These birds generally feed in small flocks of a dozen or so individuals. They form a line and herd fish into shallow water. Sometimes the group members move as a unit across the water, beating their wings or swimming with their bills submerged. Once in shallow water, the birds form a semicircle around their prey to block the fish from escaping. Then, using their large gular pouches as dip nets, the birds scoop up the corralled fish.

FISH AREN'T THE only prey that can be captured most efficiently by animals working together. Insects are another. Some social animals ensnare or flush insects.

Although most spiders live alone, at least 35 of the 35,000 known species of spiders live in groups. The most social of these spiders spin communal webs and join together to capture prey. George Uetz found that in one tiny (less than ¼-inch) species of Mexican social spider, thousands of individuals—sometimes more than 20,000—live in colonies. All members help in spinning the web, which resembles the "tent" of tent caterpillars. The web mass, constructed around the branches of small-leafed oak trees, is honeycombed with holes that provide communal retreats for the spiders. The web's outer surface, covered with sticky silk, captures prey. Spiders generally hunker down inside the web during the day, avoiding the heat. They emerge at night to spin silk and repair the web.

Whenever a fly or other prey lands on the web, day or night, nearby spiders orient toward the victim. The fly struggles, buzzes, and soon becomes entrapped in the sticky silk. Its buzzing sets up vibrations that alert additional spiders. The first spiders to reach their victim grab its legs or wings, insert their fangs, and inject venom. Other spiders bite the prey's head, thorax, and abdomen. Within 30 seconds, the spiders subdue the fly by a combination of venom and restraint. Such cooperative behavior allows the spiders to obtain larger prey than if they were solitary predators. Typically 3 to 8 spiders carry out the attack, yet as many as 20 feast on the prey.

How might such seemingly "gracious" behavior—attackers sharing with spiders-come-lately—have evolved? Spiders feed by regurgitating digestive juices into their prey. As the tissues dissolve, the spiders suck up the fluids. Uetz and his students experimented with these Mexican social spiders and found that individuals that fed in groups got more food than those that fed alone. When multiple spiders regurgitate digestive enzymes onto a prey, the tissues presumably dissolve faster, leading to a juicier slurp-fest.

Rather than clean house, these spiders leave prey carcasses in their webs. Mold colonizes the corpses and exudes an odor that attracts more flies, which become more prey for these social spiders. According to Uetz, "The fly attraction and trapping properties of *Mallos* webs have been known for many years by the Indian residents of Michoacan, who named this spider *el mosquero*, the fly eater. Prior to the widespread use of insecticides, the Indians brought web-covered branches into their homes as flypaper."

Instead of ensnaring insects, army ants flush them. Worker army ants march in columns or swarms flushing up food—just about anything they encounter, including scorpions and lizards, but especially insects. They sling their booty under their bodies and between their legs and carry the victims— whole or in pieces—back to the nest. The best-studied army ant species, the swarm raider *Eciton burchelli* common from Mexico to Paraguay, marches in raiding parties of up to 200,000 workers. Ants at the swarm front, a continuous carpet of individuals sometimes over 30 feet wide, capture the prey. The raiding group remains connected to the nest by a principal trail over which food-laden workers return to the nest and outbound ants return to the swarm. These trails can be superhighways, twelve ants wide.

Years ago when I worked in Brazil, I watched swarms of these army ants and marveled at how fast they ran over uneven ground—over twigs, around branches, and across gaps in the surface. Now we know that the reason the ants aren't stumbling into "potholes" along their highways is because other workers fill the gaps with their own bodies, providing a more level surface for food-laden nestmates.

Scott Powell and Nigel Franks studied this behavior on Barro Colorado Island, Panama. By placing an experimental apparatus with holes onto principal trails, they found that small ants filled in small holes and large ants filled in large holes. Several workers cooperated to fill in holes larger than the largest ants' bodies. When an unladen ant arrived at a hole, it first crossed the hole and spread its legs while rocking back and forth. If a good fit, the ant stood motionless except for waving its antennae. If the ant was too big, it continued on its way. If the ant was too small to cross or to fill the hole, it either waited until a bigger ant plugged the hole or it found an alternate route. Workers filled holes in less than 30 seconds, and they stayed plugged in as long as other ants continued to run over them. If no traffic passed in five seconds, the ants left the holes and continued on their way.

The workers' roadwork keeps ant traffic at a steady speed and allows food-laden workers returning to the nest to run as fast as their six little legs can carry them. It also means that the colony as a whole can bring in more prey in a day than if they were slowed down by stumbling into potholes.

ALTHOUGH MOST CARNIVORES are solitary hunters, large social mammals that eat other large mammals often hunt in groups. One pack hunter is the wolf. A wolf pack usually numbers between five and eight individuals, though some packs include eighteen or more. Most individuals within a pack are related, but, especially in large packs, some may be unrelated. When a wolf sights a  potential prey, the pack members stalk and chase it. The alpha male, leader of the pack, usually leads the chase. All members of a pack generally participate in the hunt.

It has long been assumed that pack hunting allows wolves to take down prey much larger than themselves, including caribou, elk, mountain sheep, and moose. A moose not only weighs about ten times more than a single wolf, but also can often outrun wolves. If the pack catches up with the moose, the more experienced wolves close in from opposite sides. They bite the muscles of the moose's upper leg. The victim soon stumbles to the ground, the wolves attack its throat, and the fight ends.

In fact, however, a lone wolf can capture a moose. Furthermore, field studies have shown that the per-capita gross rate of prey acquisition (measured as weight of food acquired per wolf per day) is higher in smaller packs. For example, a 27-year study carried out by John Vucetich and his colleagues on Isle Royale, an island in Lake Superior, revealed that gross per-capita prey acquisition was highest for wolves hunting in pairs. At larger pack size, the per-capita gross rate fell below that of a lone wolf hunter. So we're left with the question of why wolves commonly hunt in packs of six or more individuals instead of in pairs.

One explanation for pack hunting in wolves is kin-directed altruism—parents help their dependent young. But this can't be the only reason, because packs often include mature wolves that are not related to younger packmates. Vucetich and colleagues pointed out that biologists have long overlooked a key feature of wolf hunting ecology in trying to explain pack hunting: wolves don't get to eat all of the prey they capture because scavenging ravens steal a good portion of it.

Typically, 6 to 25 ravens hang around and eat from a wolf-killed prey carcass, but groups of more than 100 have been documented. On Isle Royale, ravens were present at virtually every wolf-killed moose carcass documented over a period of 32 years. An individual raven can eat and hoard nearly 4½ pounds of food each day from the carcass of a large prey. Multiply this by a large group of ravens, and you can see that wolves lose a lot of food. Vucetich and colleagues conservatively estimated that a pack of wolves feeding on a carcass routinely loses 4½ to 44 pounds of food per day to ravens. For this reason, wolves must kill more prey than they would in the absence of these thieves. This extra hunting costs them time and energy and exposes them to danger—cracked ribs and skulls and even death from being kicked about by large prey.

As long as ravens aren't too numerous, wolves chase these thieves away from carcasses while the wolves are actively eating or when they are resting nearby. Significantly, thievery by ravens decreases with larger wolf pack size. Another crucial part of the equation is that larger wolf packs make more kills per day on average.

Vucetich and colleagues did a series of calculations taking all of these factors into account and determined that if it weren't for scavenging ravens, wolves would be expected to hunt in pairs. But given the impact of scavengers, the costs of sharing food among more individuals in a large pack are more than offset by the smaller losses to scavenging ravens and the benefit of more frequent prey captures. The amount of food actually available per individual turns out to be higher in larger packs.

African wild dogs, about the size of German shepherds, live in packs from a few to a dozen or so individuals. The fur of these wild dogs, often called "painted wolves," is a patchwork of dark brown, yellow, black, and snow white. Like our domesticated dogs, African wild dogs communicate with body language. Their large, rounded ears stand straight up, flap, or lie flat depending on mood and intent, and their white-tipped tails wag enthusiastically—or not.

Pack hunting enables wild dogs to kill animals much larger than themselves, such as wildebeests and zebras. The pack leader selects the target and leads the other pack members to it. Within 30 minutes the dogs have usually outrun and seized the victim. All the pack members rush in and savagely begin to disembowel the prey. Subduing an adult wildebeest is dangerous business and requires teamwork. One dog grabs at the tail or a hind leg until the wildebeest slows, then a second dog sinks its teeth into the victim's lips and nose. The dogs are so efficient that a pack can kill and eat a Thomson's gazelle within ten minutes. A wildebeest or zebra might take an hour. The dogs must eat quickly or risk losing their meal to spotted hyenas. As with wolves, pack hunting might reduce the loss of food to scavengers, in this case hyenas. Unlike wolves, adult African wild dogs always allow younger dogs to eat first. After the pack members eat their fill, they return to the den, where they regurgitate food from their bloated stomachs for the pups, the pups' mother, and sick or crippled pack members left behind.

FROM BUBBLE-BLOWING WHALES to pack-hunting wild dogs, cooperation in catching dinner allows social animals to forage more efficiently and may allow them to take down larger prey than if they were solitary hunters. We humans are extraordinarily social, so it's not surprising that group hunting is part of our nature. Prehistoric peoples who banded together to trap and kill horses, bison, and mammoths must have been more successful hunters than were the unsociable solo hunters. Today we still hunt and fish in groups. We cooperate to herd, ensnare, and flush our prey. We hunt in packs. Filling potholes with one's own bodies to increase the group's daily food intake, though, is best left to army ants.

THE BABYSITTERS' CLUB

I once spent more time writing a note of instructions to a babysitter than I did on my first book.

ERMA BOMBECK, *Motherhood, the Second Oldest Profession*

Most of us probably remember (or will remember) the first time we left our firstborn with a babysitter. In my case, my husband and I entrusted one-month-old Karen to Gene, one of Pete's graduate students. Before leaving, we told Gene everything we thought he needed to know to make it through two hours. Pete and I ate dinner at our favorite Italian restaurant, thankful for the brief respite but wondering if Karen and Gene would survive the experience. We returned home to find Gene pacing the hall, holding a screaming Karen at arm's length. Karen needed a diaper change, and Gene needed a beer. Otherwise, our instructions had sufficed.

On the flip side, I remember well my teenage experiences of babysitting and being given those detailed instructions. The nervous mother who told me exactly how to brush her kids' teeth. The father who asked me to supervise his eight-year-old son's sit-ups. The parents who wanted me to take their toddler for a 20-minute walk before I fed her a premeasured amount of precooked macaroni and cheese. After dinner I was to play with her in the backyard for ten minutes. Then to bed at precisely 7:40.

Although humans make abundant use of babysitters, we're not the only ones. Some other animals care for young that are not their own, a behavior called "alloparental care." The difference is that—as far as we know—non-human animals don't give instructions to their babysitters.

AN ESTIMATED 150 SPECIES of birds feed chicks that are not their own. One is the emperor penguin, star of the 2005 movie *March of the Penguins*. These largest of all penguins weigh up to 90 pounds and breed on sea ice during the Antarctic winter. The female rolls her one-pound egg onto her feet and covers it with the lower fatty part of her belly. She soon passes the egg over to her mate's feet and waddles back to the open sea to replenish her energy reserves. Papa penguin incubates the egg for the next two months. The egg hatches just before or soon after his mate returns. To feed her chick, the mother regurgitates undigested fish, squid, and crustaceans held in an internal pouch off her esophagus. From then on, mother and father alternate between foraging at sea and staying with the chick.

Pierre Jouventin and his colleagues studied adoption in an emperor penguin colony at Adélie Land, Antarctica. Within this colony of 2,600 breeding pairs, the investigators found that some "reproductively unemployed" birds, usually females—subadults, non-breeding unmated birds, failed breeders, and non-breeding pairs—adopted chicks from the rookery. They brooded, fed, and protected one- to two-month-old chicks for as long as ten days. Half of these foster parents adopted seemingly abandoned chicks wandering about in the rookery. The other half kidnapped chicks, often following struggles with their biological parents. Why would birds adopt abandoned chicks or steal others' chicks? The answer might be hormones. In another penguin species, king penguins, females brood unattended chicks. Chick-less females retain high levels of the hormone prolactin in their blood for some time after they lose their own eggs or chicks, keeping them sensitive to begging by chicks, whoever these might be. Perhaps the same thing happens in emperor penguins. Jouventin and his colleagues suggested that if a female returning after the male's first shift can't find her mate, she might be "socially stimulated" to kidnap. Thus, maternal instinct might drive a frantic female to snatch someone else's chick—"the devil made me do it" excuse. Perhaps she gains needed experience from temporarily adopting a chick. What about the chicks, though? Adopted chicks need several short-term babysitters in succession since their foster parents don't hang around for long. Otherwise they won't survive.

Florida scrub jays provide more care for young that are not their own than do emperor penguins. These blue and pale gray-brown birds live only in central Florida, especially where oaks are mixed with saw palmetto. I once spent some time at the Archbold Biological Station and watched small groups of these jays as they went about their daily activities. I asked a researcher at the station why these birds hung out in groups. His answer opened up a whole new world of avian behavior to me: bird babysitters.

The researcher explained that Florida scrub jays breed cooperatively and have "helpers at the nest." A breeding pair of scrub jays has up to six non-breeding adults in its territory. Although physiologically capable of breeding, these additional birds instead help to defend the territory against intruders, protect nestlings against snakes and other predators, and feed the nestlings. The number of young fledged by pairs with helpers is slightly higher than for pairs without helpers. There's some benefit for the breeding pair, but why do the helpers help?

Glen Woolfenden and John Fitzpatrick carried out long-term studies and found that helpers are offspring of the mated pair. Helpers benefit from experience: when they eventually breed, they are better parents than they would be otherwise. The behavior makes sense evolutionarily and is maintained because by increasing survivorship of younger siblings with whom they share many genes, helpers indirectly accomplish the goal of reproduction: to get their genes into the next generation. These "big brothers and big sisters" often don't have the option to breed right away themselves, because all suitable habitat is taken up by breeding pairs. Unable to secure breeding territories, helpers mark time. They also compete among themselves for dominance. Top-ranking helpers will most quickly slip into vacancies when a parent bird or a breeder in an adjacent territory disappears and none of its own helpers takes over.

Some birds—for example, ostriches—provide complete care for young that are not their own. Lewis Hurxthal, in his study of Masai ostriches in Nairobi National Park, found that in May or June males establish territories up to one square mile in area. They defend these territories against other males and prepare nest sites in the sand within their territories. Females roam more widely over areas of five square miles or so, each encompassing the territories of five to seven males.

By July females begin to mate with territorial males, but only about one-third find permanent mates. These females, called "major hens," form pair bonds and sometimes stay with their cocks for many years. In contrast, "minor hens" mate with many males during the season and lay their eggs in the nests of established pairs. Each male's nest has his major hen's eggs plus those of about ten minor hens, for an average of 40 eggs per nest.

Cocks and major hens incubate the eggs for about six weeks, often threatened by jackals, hyenas, lions, cheetahs, leopards, humans, and other predators. An ostrich nest full of eggs is quite a find for a hungry predator, since each egg provides the contents of 24 chicken eggs. For a human, a nest of 40 ostrich eggs would provide the equivalent of two scrambled chicken eggs for each member in a six-person family for 80 days.

After the chicks hatch, the male and his major hen lead the young away from the nest. They protect all the chicks from predators and guide them to edible plants. Eventually these escorted broods from various nests run into each other. At that point, the most dominant parents round up other broods to join their own. By February, two or three of the original parents shepherd up to 100 chicks. Groups stay together until the following July, when the parents begin to mate anew.

Why might a major hen incubate other females' eggs, and why does a cock incubate eggs fertilized by other males? Why do adults risk their lives defending groups of chicks, only a few of which are their own? Hurxthal proposed a multifaceted answer: The system works because pairs whose nests include eggs laid by minor hens are more likely to breed successfully and pass on their genes than individuals whose nests include only their own eggs.

First, the sex ratio of breeding ostriches is skewed toward females. Predation is much higher on adult cocks than hens, presumably because the males' gaudy black-and-white plumage is more conspicuous than the females' camouflaged brown-and-gray feathers. By breeding age, therefore, hens in a population far outnumber cocks. Another factor skewing the sex ratio is that hens mature more quickly, resulting in more females ready to mate at the beginning of breeding season. Furthermore, males must establish territories before they can mate successfully; some males wait years for the opportunity to do so. What this all means is that there aren't enough males to go around for all the females.

Second, an ostrich can incubate more eggs than one hen can lay. Although an average ostrich egg weighs three pounds, the eggs are small relative to the female's body size. Whereas most small birds lay eggs that weigh about 10 percent of their body weight, an ostrich egg weighs about 1.5 percent of a female's weight. A major hen lays about seven eggs, typically one egg every other day for two weeks. Yet she is large enough to cover and incubate 21 eggs. Within several days after a major hen begins to lay, minor hens discover the nest and add to it. Prior to beginning incubation, the major hen places about 21 eggs in the center of the nest and moves the rest about three feet away. She incubates the chosen eggs during the day; the cock incubates at night. The abandoned eggs will die.

By marking eggs and identifying individual hens, another investigator discovered that a major hen recognizes her own eggs and includes them all in the incubation batch. That leaves fourteen or so eggs that she incubates that are not her own. From the minor hen's viewpoint, even though some of her eggs are likely to be abandoned, at least some probably will hatch. Thus minor hens gain with no pain. The cocks, which have "girlfriends" on the side, also benefit from this arrangement because they likely fertilized some of the minor hens' eggs as well as those of the major hen.

Third, chick survivorship is low. Hurxthal found that 90 percent of ostrich chicks within a 152-member group died within their first year. Constant nest guarding is critical because predators lurk nearby waiting to have ostrich eggs for breakfast, lunch, and dinner. Hurxthal concluded that this heavy predation pressure also gives a positive spin to incubating minor hens' eggs, from the viewpoint of the nesting pair. Think of it this way. If a nest held only the 7 eggs of the pair themselves, and a predator took 3, 3/7 of the parents' potential offspring would be lost. But if the 7 eggs are scattered among 14 eggs laid by minor hens, and a predator takes 3, on average only 1 belongs to the pair themselves and they might still raise 6 offspring. By caring for minor hens' eggs, the cock and his mate are actually protecting their own. The same rationale applies to adults that care for large groups of chicks. As always, the name of the game is to get your genes into the next generation. If you have to raise someone else's offspring to increase survival of your own, so be it.

MAMMALS ARE THE only animals that provide their young with milk. Whether low-fat, as in black rhinos' milk at 0.2 percent, or high-fat, as in hooded seals' at 61 percent, milk provides nutrition, hormones and growth factors, and antibodies that confer immunity against various diseases. Mothers of at least 68 species, from bats and rodents to humans, nurse offspring that are not their own.

Human wet nurses have practiced their trade for a long time. In ancient Egypt, women who provided this service ranged from slaves to respected and valued harem members. In Europe by the second century A.D., wet nursing was an organized commercial business. Wet nurses hired themselves out by gathering in Rome's vegetable markets at designated columns called *lactaria*. By medieval times, wet nurses—enslaved, indentured, or paid—fed other women's babies throughout Europe.

Nursing isn't the only service mammalian babysitters offer. Jon Rood, during a long-term study of dwarf mongooses from the Serengeti National Park in Tanzania, found that these reddish-brown carnivores have a complex social system involving lifelong pair bonds, communal breeding, and extensive babysitting. Packs typically consist of six to twelve members: a breeding pair, subordinate adults, yearlings, and juveniles. Only the alpha male and female, usually the oldest animals in the pack, breed. The pair retains its breeding tenure for life.

Young adult mongooses, male or female, become breeders either by staying at home and waiting for the breeders to die, or by leaving. If they leave home, they either join an existing pack or combine with other transients and form a new pack. Pack members take turns perching on termite mounds on the lookout for predators. Those not on sentry duty either rest or forage for dung beetles or other prey. When the sentry perceives danger, it utters a loud alarm call, and all the mongooses run for cover.

Every morning while most pack members forage, one or more individuals stay behind to guard the young of the alpha male and female. These baby-sitters are vigilant, fiercely protective, and will chase away predators. If the young are playing outside the den when danger approaches, the babysitter grasps the little ones behind their heads and disappears into the den. Mongooses rotate the chores so that everyone can forage. The mother usually spends the least amount of time with her young, giving her maximum time to forage and enabling her to produce more milk. Once the babies begin to eat solid food, pack members bring them insects. Although subordinate females don't breed, they sometimes spontaneously lactate and nurse the young.

It makes sense that related individuals cooperate in caring for the young. It's the same rationale as the scrub jay helpers at the nest: by helping close relatives improve the chance that their young will survive, the helper perpetuates its own genes. Thus the behavior is maintained evolutionarily. Recall, though, that sometimes young from outside have joined the mongoose pack. Rood found that many of these pack immigrants, unrelated to the alpha male and female, were devoted and sometimes even superior helpers. He suggested that reciprocity explains this behavior. Someday the immigrants might be able to breed within the pack. If so, the unrelated young mongooses they have helped feed and protect might someday help them by babysitting their young and providing sentry duty.

Female elephants are other expert babysitters. Males typically roam alone or form loose bachelor herds, but females live in tightly knit matriarchal societies of up to ten or fifteen individuals—grandmothers, mothers, daughters, sisters, aunts, and their dependent young. African elephants frequently babysit within their family groups, though it rarely includes nursing another's young. When it does, the nursing is probably only for comfort because elephant allomothers don't produce milk.

Instead, they protect the young, walking close to the babies as the herd travels. Calves wail with deep rumbles or loud bellows when distressed, frightened, or too far from their mothers. When family group members hear a distress call, they rush to the baby's aid. Allomothers rescue babies that have fallen into holes, tripped over logs, or gotten stuck in the mud. They help babies find their mothers and protect babies frightened by predators or harassed by other elephants.

Allomothering presumably establishes close relationships and cooperation among females. Thus, the behavior enhances group stability. In addition, babysitters free mothers to spend more time foraging and thus able to produce more milk for their young. Babysitting experience no doubt improves a young female's mothering ability, plus being part of the group means that she will get babysitting help once she matures and has a calf.

THE NEXT TIME you hire a babysitter, think about the elephant aunt who rescues her nephew from a mud hole. The dwarf mongoose who chases a hungry snake away from her sister's babies. The father ostrich who kicks a lion attempting to grab a few eggs for breakfast, even though the bird doesn't know if he fathered those eggs. All these animals go about their babysitting chores without instructions. Returning to the idea behind Bombeck's quote . . . maybe we shouldn't worry so much about telling the sitter which kid will want mustard, which will want ketchup, and exactly how many minutes to microwave the precooked hot dogs. A human who has survived to "teenager-dom" has probably learned enough life skills to keep children reasonably safe and well-fed for a few hours. About that dirty diaper, though . . .

SOUND THE ALARM!

Paul Revere's Ride

He said to his friend, "If the British march
By land or sea from the town to-night,

Hang a lantern aloft in the belfry arch
Of the North Church tower as a signal light,—
One if by land, and two if by sea;
And I on the opposite shore will be,
Ready to ride and spread the alarm
Through every Middlesex village and farm,
For the country folk to be up and to arm."

HENRY WADSWORTH LONGFELLOW, "Paul Revere's Ride"

On April 18, 1775, the American patriot Paul Revere spread the alarm to Lexington, Massachusetts, that the British were coming. The following day British redcoats and American minutemen fought at Lexington and nearby Concord, the beginning of the Revolutionary War.

Other social animals cooperate by sounding alarms and banding together for protection. Social defense reaches its pinnacle among termites and ants. Most worker ants use chemical alarm signals to alert nestmates of danger. In some ant species, these same chemicals also serve to recruit those nestmates for colony defense. Many ant colonies have specialized workers, the soldiers, which sting, bite, or rip their enemies in two with shearing mandibles to protect the colony. Soldier termites of some species secrete chemicals from glands on their heads; these chemicals appear to serve as alarm signals. Some soldier termites have huge, powerful mandibles. Others have needle-sharp, snapping mandibles that plunge deeply into flesh. My favorite: soldiers of nasute termites manufacture and store a defensive material in a Pinocchio-type structure called a "nasus" sticking out from the front of the head. When confronted with danger, a soldier points its nasus at the subject, contracts its head muscles, and sprays liquid that quickly thickens to a glue-like consistency and gums up the enemy. The nasute soldier's marksmanship is amazing, especially considering it is blind. Unable to reload quickly, after shooting its wad the soldier wipes its nasus on the ground and retreats inside the nest. Other soldiers take up the battle, attracted by alarm substances in the spray.

VERTEBRATES LIVING IN social groups also protect each other through behaviors such as mobbing, alarm signals, and sentinels.

If you've ever approached a colony of nesting seagulls and incurred their wrath, you know that mobbing behavior, the gathering of individuals around a potential predator, is effective. Just about any predator—or a well-meaning bird-watcher—will turn and flee when dive-bombed by hundreds of screaming, defecating birds.

Some colonially nesting fishes also mob. Wallace Dominey studied nesting bluegill sunfish in Cazenovia Lake, New York. There, males constructed bowl-shaped depressions for nests in tightly packed colonies of up to 500 individuals. Males competed with each other for the chance to mate during a one- to two-day spawning period. Those that successfully mated and ended up with fertilized eggs in the nest then defended their nests from predators over the next eight to ten days. Dominey released a snapping turtle—a natural predator of bluegills—into several spawning colonies and watched the bluegills' reactions. As the turtle cruised through a colony, both males and gravid females rapidly approached, from behind so as to avoid the turtle's jaws, and followed the turtle until it left the area. When Dominey repeated the experiment with a painted turtle, incapable of preying on bluegills, the fish did not mob the intruder.

Bluegills aren't the only colonially nesting fish that mob. In coral reef areas, domino damselfishes mob octopuses and predatory fishes such as barracudas that approach their nesting sites. Some fishes initiate mobbing after one of their own is captured. Dominey observed a school of young bluegill mob a chain pickerel, a fish with a large mouth and big appetite. After the pickerel captured one bluegill, several other bluegills swam to within two inches of the predator's mouth. The pickerel retreated about three feet, bluegills following close behind, and then rapidly swam away. Another time, Dominey watched a Florida softshell turtle enter the nest of a black bass, where it chomped on mud and presumably ate some eggs from the nest. Several other bass nesting six feet away ganged up, bit and shook the turtle's tail and legs, and drove the turtle from the nesting area.

MY LONG-HAIRED DACHSHUND'S favorite hiking spots are high in the San Juan Mountains of southern Colorado, where he races through meadows of columbine and paintbrush terrorizing the yellow-bellied marmots. The instant a marmot spots Conan, it whistles and disappears down its burrow. Conan never seems to tire of the game.

Since 1962 Ken Armitage and his students and colleagues have studied yellow-bellied marmots in and around the Rocky Mountain Biological

Laboratory a bit farther north in Gothic, Colorado. They have discovered that marmots live in groups made up of a breeding female and her older and younger offspring. Adult males immigrate into areas that have one to several social groups. When alarmed, marmots dash to the entrance of their burrows and often whistle—mouths quivering and bodies shaking— just before they disappear inside. Armitage and his colleagues suggested that marmots give alarm calls primarily to warn their off-spring of danger. Before the year's pups had emerged from their natal burrows, the investigators found no significant difference in the frequency of alarm calling among adult female mothers, adult females without offspring, adult males, yearling females, or yearling males. Once the pups were above-ground, however, mothers called significantly more often than did mar-mots of any of the other age or sex classes. Since Conan and I usually hike in Colorado in late August, most of the whistling is probably from protec-tive mother marmots warning their pups: "Get inside, kids!"

MANY OTHER MAMMALS and many birds that live in social groups give alarm calls. Some of these, such as vervet monkeys and Gunnison's prairie dogs, give specific calls for specific predators. Many species that give alarm calls also post sentinels that stand guard and alert other group members of impending danger. The advantage is that all but the sentinel can devote more time to other activities such as food-gathering without losing protection from predators, as in the dwarf mongooses highlighted in the previous essay.

Some vertebrates display all three behaviors: sentinels, alarm calls, and mobbing. Recall that Florida scrub jays, the "helpers at the nest" spe-cies from the previous essay, live in groups made up of one breeding pair and several "helpers," yearling or older offspring that hang around and defend the territory, protect against predators, and help feed their younger brothers and sisters. Kevin McGowan and Glen Woolfenden observed that family groups of Florida scrub jays have a sentinel system in which one bird stands guard at a time. Breeding males perform sentinel duty more often than do their mates. The sentinel typically sits on an exposed perch, constantly turning its head slowly to one side then to the other, scanning its surroundings. If the sentinel spies a potential predator, it gives an alarm call and the others respond. If the danger is a flying raptor, such as a Cooper's hawk or sharp-shinned hawk, the birds either watch alertly, ready to flee, or they hide. If the predator is on the ground, they mob it. Florida scrub jays commonly mob snakes, especially coachwhips and indigo snakes but also coral snakes and rattlesnakes. The birds mob alligators, gray foxes, dogs, raccoons, longtail weasels, bobcats, house cats, and humans.

Ann Francis and two collaborators conducted experiments in which they released a live Florida pine snake in the presence of Florida scrub jay groups, then watched the birds' responses. In each of 52 trials, the jays mobbed the snake. In every case except one, the bird that first saw the snake began the mobbing. First the bird uttered a harsh, scolding call and fixed on the predator while in full view of it. The bird then did one or more of the following: approached the predator, flicked its wings or tail, jumped, bobbed, pecked its perch, and/or bit the snake's tail. In most cases, all group members mobbed. Breeding males mobbed for longer and approached snakes more closely than did female breeders and helpers. Only breeding males attacked and bit the snake. Fledglings fewer than 47 days old didn't join in the mobbing, presumably because they were still klutzy fliers.

How effective is mobbing? In 31 of the 52 trials, the snake remained motionless except for occasionally flicking its tongue. In the remaining 21 trials, it slithered away. Mobbing by scrub jays probably makes it harder for snakes and other predators to capture a relative.

YOU'VE NO DOUBT seen photos of meerkats—comic-looking mammals that stand upright on their back legs as they scan for predators. These small burrowing creatures, a species of mongoose, live in groups of up to 50 in semi-desert regions of southern Africa. As with the dwarf mongoose featured in the previous essay, members of meerkat groups all help raise the young and guard against predators. Large eagles, jackals, snakes, and other predators prey heavily on meerkats. The meerkats' defense is a sentinel system. Group members rotate being on guard duty. Sentinels belt out alarm calls whose acoustic structure, frequency of occurrence, and other characteristics relay information on type of predator and urgency. If a raptor flies overhead, meerkats give the most urgent call, the "panic call," once or twice in very short intervals. Everyone runs pell-mell into the burrow.

Terrestrial predators on meerkats elicit other types of calls that induce group members either to move away or to mob the predator. Mobbing behavior typically begins with one or more meerkats, tails erect and heads bobbing, approaching the animal while calling. Their calls recruit other group members to join in.

COORDINATED VIGILANCE and group defense by humans have come a long way since Paul Revere's famous ride. We join Neighborhood Block Watch crime prevention programs. College campuses use Emergency Text Notification to spread word of rapists and deranged shooters. High-tech security systems sound alarms and alert the police if anyone breaks into our homes. As I wrote the above sentences, a thought dawned on me: The difference between most of our protection systems and those of other animals is that ours protect us from other people, our main predators. For other social animals that sound the alarm for group members, the predator is a different species.

AN INTIMATE ACT

Sharing food with another human being is an intimate
act that should not be indulged in lightly.

MARY FRANCES K. FISHER

Sharing food is indeed an intimate act. As a child, I sat on Dad's lap and begged for the funny-flavored green olive from his martini. At lunchtime I ate my younger sister's vegetable soup when Mom wasn't watching—in exchange for a future favor. Jan hated soup with pieces of anything in it. Vegetable soup was the pits. Now I give the crispy skin off my Thanksgiving turkey to my husband, Pete. He enjoys it more than I do. Before paranoia set in over the past couple of decades, schoolchildren shared their lunchbox contents with one another. Adults share food at potlucks and neighborhood barbeques. But most humans don't carry food sharing to the extreme of ants and vampire bats, which regurgitate for each other.

LET'S START WITH ANTS. To understand food sharing in ants, we need to examine the insects' anatomy and their complex social system. As a group, ants eat both liquid and solid (or at least squishy) food. Liquid food includes nectar and honeydew (sugary liquid excrement that certain sap-tapping insects such as aphids produce). Fungi, seeds, fruits, flowers, and other animals such as insects and spiders make up the hard or squishy items. But ants can't ingest solid food, only liquids. So what can they do? A typical ant chews a solid food item with its maxillae, located behind the mandibles, and breaks it up into smaller particles. The particles then pass into a pouch beneath the worker's tongue, where muscles contract and squeeze out the liquid. The ant spits out most of the compacted solid matter as a pellet and swallows the liquid. The liquid passes into a distensible pouch, the crop, where it is stored undigested. From time to time, the crop contracts and pumps liquid food into the stomach for digestion: the worker ant's personal food supply. The rest of the liquid stays in the crop for communal use. More about this shortly.

Most ant colonies include three castes: the queen(s), males, and, most abundant of all, the non-reproductive female workers. The males' sole job is fertilizing young queens, which then spend their lifetimes laying eggs. The workers care for the queen and the young, repair and defend the nest, and gather and distribute food throughout the colony. Each worker has her own job. Since not all workers gather food, foragers must share food with their nestmates, which stay behind and perform other chores. Ants do this in diverse ways, depending on the species.

Some members of the relatively primitive ant subfamily Ponerinae distribute food to colony members through a "social bucket" system. Workers collect liquid food such as honeydew and fruit droplets and carry it between their mandibles back to the nest. When hungry ants tap the donor's head and mandibles with their antennae, the donor transfers part of her droplet into the gaping mandibles of the nestmate. Ten or more ants may share one worker's booty in this crude food-sharing system.

Workers of most ant species, though, use a more refined type of food sharing called "oral trophallaxis"—they regurgitate liquid food from their crops for nestmates. The crop serves as a "social stomach." A nestmate lightly taps a worker ant with her antennae or forelegs. The worker turns to face the tapper. If the latter continues to tap on the worker's lower mouth, the donor regurgitates a droplet of her crop liquid.

Honeypot ants from the deserts of the New and Old World take food sharing to extremes. Foraging worker honeypot ants gather termites, plant nectar, and honeydew. They return to the nest and feed their bounty to large workers called "repletes." The liquid food, stored in the crop, swells the repletes' abdomens to the size of peas. They're so bloated they can hardly walk.

The repletes then climb to the chamber ceiling and hang on by their claws, where they serve as live storage tanks for the colony. When food is scarce, nestmates tap the repletes with their antennae and the repletes regurgitate sweet syrup. A colony of 15,000 honeypots maintains about 2,000 of these living storage tanks. Repletes are so valuable that when honeypot colonies raid smaller colonies, in addition to stealing larvae and pupae for slaves, they drag away repletes for their own pantries.

We know less about the function of anal trophallaxis. In some kinds of ants, larvae secrete a milky liquid from their anus. In a strange twist—larvae feeding adults—workers slurp up the droplets, which presumably serve as supplemental food. Workers of certain other ants consume droplets of rectal fluid from each other. In one species, newly emerged workers solicit droplets from older workers by licking the tips of their abdomens. Given that young termites (unrelated to ants) get their cellulose-digesting bacteria and protozoans by eating adult termites' feces, might these young worker ants gain beneficial microbes from their elders? And then there's a species of slavemaker ant in which the workers and queens occasionally extrude droplets of anal liquid, which the slaves eat—a rare example of social parasites giving something to their captives. Is this food or merely some sort of dominance mechanism? We don't know.

Some ants share food when it's still solid. As mentioned in the essay "Bubble Blowers," army ants flush up their prey. Up to a half million workers march in columns or swarms flushing up tarantulas, scorpions, grasshoppers, cockroaches, beetles, and small vertebrates, including small snakes and lizards and nestling birds. Successful hunters sling their booty, either whole animals or fragments, under their bodies and between their legs, then march back to the bivouac site, where hungry nestmates feast. Another solid food sharer is the harvester ant. Forager harvester ants carry seeds back to their nests, where they store the food in chambers. When hungry, workers tear off the outer seed coatings, chew the "meat," swallow the liquid, and spit out the solid. All the workers share the seeds.

Humans have learned to destroy certain pest ants by turning ants' natural food-sharing behavior against them. When workers pick up food containing insecticide, they later regurgitate the poison from ant to ant throughout the colony, including the reproductive queen(s). The social behavior of sharing food that could someday allow ants to rule the world might also be their undoing.

NOW LET'S TURN to "warm-blooded" animals—in all senses of the term. Common vampire bats feed on the blood of other mammals, either on livestock such as cattle, horses, and pigs, or wild animals including deer, peccaries, and tapirs. Occasionally even on an unfortunate human. Getting that blood meal is hard work, and bats aren't always successful. After locating a sleeping victim, the three-inch bat must find a relatively bare, warm spot

with blood vessels near the skin surface, such as a nose, neck, or ear, which it locates through heat-sensitive cells in its nose. The bat then slices out a small piece of skin with its razor-sharp upper incisors. Anticoagulants in the bat's saliva keep the blood flowing while the bat laps with its tongue. A successful bat will consume over half its body weight in blood within 30 minutes. The bite causes little pain, and the victim rarely awakens. If it does wake up, though, the victim brushes off the bat. Imagine being brushed off by an animal that is thousands of times your own weight! Young bats often get the brush-off. They're klutzy and haven't yet mastered the technique of biting quickly without attracting notice. Fortunately for them, feeding skills improve with age. One- to two-year-old bats successfully feed two nights of every three. By the time they're older, they successfully feed better than nine nights of every ten.

Unsuccessful bats—mostly the youngsters—return to the roost and solicit food by licking a roostmate's lips. Gerald Wilkinson studied common vampire bats in northwestern Costa Rica. He watched bats regurgitate blood into other bats' mouths. Mothers periodically regurgitated blood for their nursing offspring, and adult females regurgitated blood for other adult females. By sharing blood meals, these bats presumably risk transmitting saliva-borne rabies virus to each other. But the risk of starving is more serious. A common vampire bat that goes three days without food will die.

Common vampire bats usually live in colonies of 20 to 100 individuals in caves or hollow trees. Within a colony, groups of females often roost together for many years, though they may use several different caves or trees within a given week. Some individuals are related, because daughters generally stay with their mothers after reaching sexual maturity. Because some females occasionally switch roosts, though, a colony also will include unrelated individuals. Wilkinson found that 70 percent of 110 instances of blood sharing took place between mothers and their young. In the other 30 percent of cases, adult females regurgitated for pups other than their own, adult females fed other adult females (both relatives and non-relatives), and two adult males fed pups. The bats didn't regurgitate for just anyone. They only did it for relatives and female roosting buddies.

Wilkinson's long-term field study suggested that adult males are on their own. No one shares blood with them, not even their roosting buddies. Later, though, other scientists carried out a seven-month study on a captive colony of common vampire bats. There, males fed each other regurgitated blood three times and females fed males three times. Clearly, we have more to learn about food sharing in vampire bats.

WHY DO ANTS and vampire bats share food? The "why" questions in biology are always the hardest to answer; often the best we can do is speculate. Only certain ants from a given colony forage for food, which they share with non-foraging nestmates. Scientists explain this altruistic food sharing in ants through kin selection: by helping their relatives, workers increase survival of their own genes. Kin selection also explains much food sharing in vampire bats, especially when mothers feed their own offspring. Wilkinson suggested that reciprocal altruism might explain bats feeding non-relatives: "You do something good for me, and someday I'll return the favor."

Of course, ants and vampire bats aren't the only friendly regurgitators. African wild dogs, wolves, mountain lions, some birds, some insects other than ants, and certain other animals regurgitate food for their young. Some even do it for mates.

Although humans don't normally regurgitate to share food, in some cultures we pre-chew food for others. No doubt prehistoric women chewed mastodon meat, roots, and berries for their babies. In some cultures, parents still chew food for their infants and toddlers. In fact, one theory on how the kiss originated revolves around sharing food: mothers pre-chewed food and passed it on to their babies in a "kiss." The smooch may then have evolved between adults as a sign of affection. In some cultures, people with a good set of choppers chew food for old and toothless members of the group. Still, as far as we know, Count Vlad Dracula didn't share his blood meals . . .

WHATEVER HAPPENED TO BABY BOOBY?

When I first put my hand through a slit in the oviduct I received the impression that I had been bitten. What I had encountered was an exceedingly active embryo which dashed about open mouthed inside the oviduct. The teeth were not strong enough to penetrate my skin but were sharp and hard enough to produce a pricking sensation.

STEWART SPRINGER, "Oviphagous Embryos of the Sand Shark, *Carcharias taurus*"

When young, or not so young, most of us squabbled with our siblings. One of my younger sisters was deathly afraid of spiders and their kin. I took advantage of her phobia and on more than one occasion terrorized her with daddy longlegs. Of course now I realize my behavior was cruel (both to Jan and the daddy longlegs), but my cruelty pales in comparison with the expression of rivalry among sibling sand sharks. These siblings kill one another. And they do this before they're even born!

The violated oviduct and its inhabitant described above belonged to a sand shark. The hand belonged to a curious scientist. One of the shark's oviducts housed a 10¼-inch pinkish-white embryo and 71 egg capsules, 10 of which were empty shells. The other oviduct held a 10½-inch pinkish

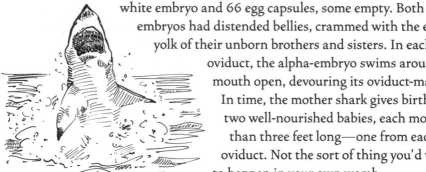

white embryo and 66 egg capsules, some empty. Both embryos had distended bellies, crammed with the egg yolk of their unborn brothers and sisters. In each oviduct, the alpha-embryo swims around, mouth open, devouring its oviduct-mates. In time, the mother shark gives birth to two well-nourished babies, each more than three feet long—one from each oviduct. Not the sort of thing you'd want to happen in your own womb.

Another animal that kills siblings before they're born has nothing shark-like about it. It's the beautiful and graceful pronghorn, the fastest-running mammal in North America. The first blastocyst (early embryonic stage consisting of a hollow ball of cells) to implant in each horn of the uterus sends out invasive processes that pierce and devour all subsequent embryos that implant in that horn. No matter how many eggs the female ovulates, no matter how many of them get fertilized, she gives birth to two babies at most. Her first two blastocysts see to that. Pronghorn embryos don't eat their siblings as do sand sharks. Rather, siblicide may persist in pronghorns because mothers cannot care for more than two young at a time.

Queen honeybees are siblicide survivors. A colony generally rears 15 to 25 presumptive queens—let's call them "princesses," although that term isn't really used—for each new generation. Only one, though, survives to become queen of that colony. Here's how it works. The old queen lays one pearly white egg in each cell of the brood nest. After the eggs hatch, workers feed the hatchling larvae from a few cells usually located along the lower margin of the comb with a special food that ensures they will develop sexually into princesses. While the princesses develop, their mother queen's relationship with her other subjects changes. She becomes restless. The workers feed her less and become mildly hostile. They may even pummel her and jump on her. Eventually they push the old queen out of the hive. She and a retinue of her subjects fly off in a swarm. In time they will found a new colony.

Meanwhile back in the original colony, the first virgin princess emerges. She wanders around and chews holes in the walls of her not-yet-emerged sisters' cells. This induces worker bees to kill those other princesses, still larvae or pupae, before they can emerge. In the rare case where two or more princesses emerge from their cells simultaneously, they try to sting each other and spray rectal fluid at their rivals. Ideally, the fluid makes the target princess less aggressive, giving the sprayer time to sting first. Most often only one princess emerges from this battle, but sometimes more than one survives.

Workers urge the surviving virgin princess or princesses out of the nest and on to her or their nuptial flights. Each may make twelve or more flights, mating with a different male each time. Honeybee sex is rough. As mentioned in the essay "Not Tonight, Honey," the male literally explodes his genitalia inside the queen and then dies. Once the new queen—no longer a virgin princess—has enough sperm to last her a lifetime, she either flies off in a swarm with workers and founds a new colony, or she returns to her natal colony, kills any remaining sisters, and takes over the nest.

MOST PERPETRATORS OF SIBLICIDE, like honeybee princesses, accomplish their dirty deeds after they are born or hatched. In many cases, cannibalism follows siblicide. That is, one sibling kills, then eats, another—or all the others. Certain spiders, snails, termites, caterpillars, fish, and tadpoles eat eggs or newly hatched young from the same clutch, which provides them with lots of food. When young develop and grow at uneven rates, faster-growing, larger individuals chow down on slower-growing, smaller siblings. Because they "eat at home," the newly hatched young don't have to wander about seeking other food. By moving around less, the young are less exposed to other animals that might eat them. Biologists call this behavior "selfish siblicide" and attribute its evolution to straightforward natural selection: by eating siblings, the young cannibal potentially increases its own genetic contribution to the population.

Sometimes it pays to grow slowly, though. I once watched what I interpreted as an age-reversed example of "selfish" cannibalism in Cuban treefrog tadpoles. Cuban treefrogs, native to Cuba and neighboring Caribbean islands, are now exotics found in Puerto Rico, several Lesser Antillean islands, Hawaii, and mainland Florida. People probably accidentally introduced these five-inch, gray, light brown, or green frogs to the lower Florida Keys in the 1800s. Since being introduced, the frogs have expanded their range throughout south Florida, into central Florida, and even up into pockets of north Florida. These invasive frogs have a huge appetite, and they eat native frogs. They also eat each other.

One of my students brought me some Cuban treefrog tadpoles he had collected from a small plastic wading pool in Fort Myers, Florida. I put them all together in one aquarium. I don't know how many females had laid eggs in that small wading pool, but quite possibly all the tadpoles were sisters and brothers. Some grew faster than others and began to metamorphose sooner. Once their front legs emerged, metamorphosing individuals climbed onto the side of the aquarium, leaving their tails and sometimes their hind legs dangling in the water. Their less developed companions chomped down on the tails or hind legs and pulled the victims back into the water. Within seconds other tadpoles joined in, attracted either by the motion of the struggling tadpoles or by chemical cues. The cannibals quickly devoured their more-advanced victims—likely their brothers and sisters. In the real world, the cannibals potentially would have increased their relative numbers of descendents by reducing the competition.

At first glance, siblicide "makes sense" in terms of natural selection. If your chance of having numerous grandchildren increases when you eat your rivals, be they siblings or not, your genes will spread in the population. Consider, though, that on average a sibling shares 50 percent of your genes. So there's a catch. Catch-22. If you eat a sibling, you're eating some of your own genes. If you eat more than two siblings, you might be decreasing, not increasing, the potential for your genes to spread. But if you show good manners and decline to eat siblings, you might die, and the genes you carry will die with you. No grandchildren.

In the evolutionary concept of inclusive fitness, the fitness of an individual depends not only on the relative number of its genes that it passes on directly to grandchildren but also the genes it passes on through relatives that are not descendants. Biologists often refer to inclusive fitness as "kin selection." Does siblicide make sense from the standpoint of kin selection? That depends on the situation. Consider the orange-and-black swamp milkweed leaf beetle.

Infancy is tough for many species of beetles. Newly hatched beetles often die before finding their first meal. In swamp milkweed leaf beetles, which lay their eggs in clutches, the early hatching larvae solve this problem by eating food lying around next to them—their unhatched or late-hatching brothers and sisters. The nutrition gives the cannibals a head start on growth and survivorship. If beetle larvae were well mannered and none ate a single sibling, all might die before finding the first meal of a juicy green leaf. If some larvae eat their brothers and sisters, though, at least the cannibals might survive and pass on the genes they possess, genes also shared with their sibling fodder. Kin selection as an explanation of siblicide makes sense in these beetles that otherwise have a poor chance of surviving.

In some cases, kin selection should reduce the occurrence of siblicide. For example, if individuals that eat their brothers and sisters leave behind less of their shared genetic makeup than do other individuals in the population that eat relatives less often, siblicide would be a disadvantage. Following this line of thought, then, we should expect that some cannibalistic species should recognize kin and avoid eating them.

One such species is the plains spadefoot toad, two-inch gray amphibians that range from southern Canada to northern Mexico. Plains spadefoots spend much of their lives underground. Adults emerge to breed in puddles and water-filled ditches following heavy rains. The tadpoles grow and develop quickly, metamorphosing in less than two weeks. Fast development is critical because these temporary bodies of water often dry quickly.

Tadpoles of the plains spadefoot toad come in two body forms, or morphs: a "typical" morph that eats primarily plant and animal detritus, and a "cannibal" morph that eats live prey, including other plains spadefoot tadpoles. David Pfennig and his colleagues offered tadpoles of the cannibal morph a choice between typical morph siblings and typical morph non-siblings. The tadpoles more frequently ate non-siblings, which they apparently identified through a chemical taste test. The cannibal morphs sucked tadpoles of both typical morph groups into their mouths equally often. If the tadpoles were siblings, the cannibal spit them out unharmed. Non-relatives continued on down the hatch. Curiously, cannibals deprived of food for 48 hours did not distinguish between relatives and non-relatives. It seems that hunger reaches a point where instinct kicks in, kin or not!

Tiger salamander larvae also come in two morphs: "typicals," which eat mainly invertebrates, and "cannibals," which have huge heads and wide mouths and eat other tiger salamander larvae. When Pfennig and his colleagues gave cannibal morph larvae a choice of siblings, cousins, and non-kin, the cannibals avoided eating siblings. When offered cousins and non-kin, the cannibals preferentially ate non-kin. How do they recognize kin? When the investigators blocked the cannibals' nostrils with an adhesive, the larvae could not distinguish between relatives and non-relatives and ate indiscriminately. The larvae apparently recognize relatives by their odors.

SIBLICIDE IN BIRDS seems to result primarily from "decisions" that parents make. That is, the behavior of some bird parents enhances the survival of stronger offspring to the detriment of others that end up as victims. Many species of hawks, buzzards, kestrels, eagles, and owls, plus some herons, egrets, and other birds, exhibit "brood reduction"—a euphemism for getting rid of siblings so Mom and Dad give the survivor all the food. How do bird parents set the scene? Most birds lay only one egg per day. If the parents wait to begin incubating until the female has laid all her eggs, they put all their young on an equal developmental footing because the eggs will all hatch at the same time. If the parents begin incubating the first egg right away before the female lays other eggs on subsequent days, the parents give their first-laid a developmental head start. In this case, chicks hatch at about the same intervals as the eggs were laid. The first-hatched will always be larger and stronger than its younger siblings. Why would parents behave so as to exacerbate sibling rivalry? Let's look at two possibilities.

Swallow-tailed kites range from the southeast United States, through Central America, and into tropical South America. The tail is deeply forked, resembling that of some swallows, thus the common name. These magnificent black-and-white birds, with a four-foot wingspan, nest in the tops of tall trees. Both parents share in raising the young . . . or rather the firstborn.

In a study of the reproductive behavior of swallow-tailed kites in northern Guatemala, Richard Gerhardt and two colleagues found that most females laid two eggs per nest, but that no nest fledged more than one chick. The investigators climbed trees to watch the sequence of events. Female kites began to incubate after laying their first eggs. Three to four days later, they laid the second eggs, which weighed significantly less than the first. In each case, the first egg produced the successful fledgling. When the second had hatched, the first-hatched pummeled its younger sibling's head and neck with its bill and shook it by the head, neck, or wing. Sometimes the attacks consisted of only a few blows; other times the first-hatched pummeled and shook its sibling for up to 30 minutes. Parents did nothing to intervene, and the second chicks never fought back.

Being more aggressive and dominant than their siblings, the first-hatched chicks got most of the food the parents brought to the nest. Second chicks died within five days after hatching. Gerhardt and his colleagues concluded that second chicks died of starvation, possibly combined with internal injuries. And a mother's reaction? One female awoke to find her second chick dead. She picked at the body, ate a few bites, then flew from the nest with the carcass and dropped it. Not what we usually think of as "motherly love"!

So why do female swallow-tailed kites lay two eggs if only the first hatchling will survive? Perhaps the first doesn't always survive after all. The second egg and chick might serve as an insurance policy. If the first egg or chick dies, the second might survive—providing it isn't too beaten up already. The cost to the female of laying that second egg is small: it is less than 8 percent of her body mass, and she doesn't spend much time or energy feeding the runt that hatches from it.

In swallow-tailed kites, food shortages apparently were not the triggers for siblicide. The first-hatched didn't beat up on the runt only at feeding time, and parents probably could provide enough food for two chicks. Blue-footed boobies are a different story.

When I saw my first blue-footed boobies on Santa Cruz, one of the Galápagos Islands, it was like seeing old friends. They looked just like the many photos I'd seen—comical, white-and-brown, nearly three-foot-long birds perched on the most amazing fully webbed bright

blue feet. The name "booby" comes from the Spanish *bobo*, meaning
"dunce" or "fool." The boobies awkwardly strutted around, looking more
than slightly tipsy. Blue-footed boobies nest in colonies. Both parents
care for the young, and at least one parent is always with the brood.
The parents regurgitate fish directly into their chicks' mouths.

Observations reported in the 1970s suggested that when food is limited, only one chick survives from the two-chick brood. To test the idea
that sibling aggression in blue-footed boobies increases with food depriva-
tion, Hugh Drummond and Cecilia Chavelas studied birds on Isla Isabel,
a small volcanic island in the Pacific Ocean off the coast of Mexico. The
investigators taped the necks of both first and second chicks in each of
41 broods so the chicks couldn't swallow regurgitated fish. Don't worry.
The soft adhesive cloth tape the investigators used did not constrict the
chicks' necks, and the chicks showed no signs of pain or physical harm.
Drummond and Chavelas compared behavior of the experimental chicks
with behavior of control chicks. These control individuals had their necks
taped only during part of the day, but the tape was removed whenever a
parent attempted to feed a chick. Thus, the control chicks experienced
some of the physical effects of having their necks taped, but could ingest
all the fish their parents regurgitated for them. Amazingly, this type of
experiment is possible with blue-footed boobies because they are so toler-
ant of people.

And the results? Control chicks behaved normally and gained weight
during the experiment. Experimental chicks lost weight. As expected,
the first chicks pecked more at their younger siblings when deprived of
food, especially once the first chicks' weights declined to 20 to 25 percent
below normal for their age. First chicks begged more often than did second
chicks. Parents responded by trying to feed the first chicks more. After the
investigators removed the tape from the chicks' necks so they could swallow food, the pecking, begging, and feeding behaviors returned to baseline
values, suggesting that the older chick is more aggressive when it doesn't
get enough food. By pecking at its younger sibling and begging more when
its parents arrive with food, the first-hatched chick gets the lion's share of
the food. If there's a severe food shortage, only the first-hatched survives.

BE GLAD YOUR brother or sister only tattled on you. Or threw daddy
longlegs on you. Maybe he beheaded your Barbie or she stole your candy,
but it could have been worse: you could have been a late-developing sand
shark embryo, the fastest-developing Cuban treefrog tadpole, or the
second-hatched blue-footed booby. Or you could have been Roman em-

peror Caracalla's younger brother. The tempestuous Caracalla and the more intellectual Geta were sons of Septimus Severus, emperor of the Roman Empire from A.D. 193–211. In an attempt to avoid inciting sibling rivalry, Severus proclaimed that upon his death the boys would both rule the Roman empire as co-emperors. Big mistake. Caracalla and Geta hated each other. When Septimus Severus died in February 211, Caracalla, 22 years old, and Geta, 21, each wanted to be the one and only emperor—no "co" about it. In December 211, instead of pummeling his younger brother directly as would a young swallow-tailed kite, just like a honeybee princess (another royal family!), Caracalla arranged for his brother to be assassinated. Caracalla ruled alone, but only until 217, when an officer in the imperial bodyguard in turn assassinated him while he was urinating by the side of a road. And that's just one example of hundreds throughout history of siblicide, direct or indirect, in human royal families.

Just for the record, I'm not from a royal family, and even though I tortured my sister Jan with daddy longlegs, I never teased my younger brother Alan . . .

2 Taken to the Cleaners, & Other Interactions between Animal Species

CAPUCHIN MONKEYS WIPE millipede secretions over their fur as insect repellent. Mites ride in hummingbird nostrils to get to their next meal of nectar. Shrimps clean dead tissue and parasites from lobsters. Parasitic wasp larvae live in, and feed on, alfalfa weevil larvae. Mosquitoes filch honeydew droplets from ants.

These are just a few examples of the many animals that "use" other species of animals in one way or another. Some relationships between species are casual opportunism, as with capuchins wiping millipedes over their fur. Others are not casual at all. They are symbiotic—associations that involve one species living in, on, or with another species in a close relationship in which at least one species benefits. In 1879 German botanist Heinrich Anton de Bary coined the word *symbiosis* from the Greek *sym,* meaning "together with," and *biosis,* meaning "way of living." Biologists categorize symbiotic interactions into three main types: commensalism, mutualism, and parasitism.

A relationship in which one party benefits and the other is unaffected is called "commensalism." For example, mites that hitch a ride on a hummingbird gain transport; they neither benefit nor harm their hosts. The word "commensalism" comes from the Latin *com,* meaning "with," and *mensa,* meaning "table," though not all commensalistic associations involve food.

If both parties benefit, the relationship is called "mutualism." Shrimps that groom lobsters get food, and the lobsters get cleaned. The word "mutualism" comes from the Latin *mutare,* "to exchange."

A relationship in which one individual benefits and the other is harmed but not killed—at least not right away—is called "parasitism." Some parasites live in or on the host. A female parasitic wasp injects her eggs into an alfalfa weevil larva. The wasp's larvae feast on their host's fluids and tissues, but don't kill it until they've finished growing. Social parasites harm other animals indirectly, for example, mosquitoes steal honeydew from ants. The word "parasitism" comes from the Greek *para*, meaning "beside," and *sitos*, meaning "food." In ancient Greece, a person who flattered and amused his host to get free food was called a *parasitos*.

Sometimes the nature of the relationship—whether only one partner gains or both gain, and whether one is harmed—is not clear. Furthermore, the relationship might change depending on environmental factors and population densities, and species interactions constantly evolve through time. The dynamic nature of these interactions in space and time provides biologists and naturalists with an unending diversity of mysteries to examine.

HUNTING PARTNERS

As I walked through the jungle, intent on birds and enjoying having a pristine wilderness entirely to myself, I became vaguely aware of a black shape somewhere behind me. I stopped, looked around, saw nothing, and walked on, slightly unnerved. Again I had a sense of a figure following me, quickly turned, and glimpsed a black creature vanishing behind a tree. I felt my heart pounding as I reflected that I had no radio to call for help and the helicopter wasn't due back for a week.

JARED DIAMOND, "Strange Traveling Companions"

Ecologist Jared Diamond spun around again and spied a drongo, a midsize black bird, in the New Guinean rain forest he was exploring. The bird meant him no harm. It was simply following Diamond to capture insects he flushed while walking. In Africa and continental Asia, drongos follow large wild mammals such as elephants and giraffes, eating the insects they flush. In New Guinea, where there are no large wild mammals, drongos follow flocks of birds. Or people.

The North American equivalent of the drongo is the cattle egret, a bird many of us know well. Originally from Africa, where they follow wild ungulates, cattle egrets dispersed to South America early in the twentieth century. These vagrants bred successfully, their populations grew, and they continued to expand their range. By the early 1950s, cattle egrets had reached the United States, where they have become common and widespread. These elegant white-plumed birds follow cattle and eat the grasshoppers and other insects they flush. Cattle egrets can find food on their own, but they capture up to 50 percent more prey by following cattle. These birds are opportunistic: they also trail behind moving tractors or other farm machines that flush insects.

DRONGOS AND CATTLE EGRETS aren't the only birds that eat insects flushed up by other foraging animals. Army ant birds are another example. At least 50 species of birds in New World tropical forests regularly follow army ants and prey on insects fleeing from the ants. As described in "Bubble Blowers," swarms of army ants march along the ground and flush insects, spiders, lizards, frogs, and other small vertebrates, some of which they sting and transport back to their bivouac for supper. "Professional" army ant–following birds get at least half, and often most, of their food from the flushed animals the ants don't catch. In many forests there's almost always an ant colony on the march, so following ants is a dependable way to get food.

A single large ant swarm can attract 30 species of ant-following birds. The larger of these dominate the prime central zones, leaving peripheral areas to smaller birds. Ground-cuckoos walk or hop along the edges of the swarm. Woodcreepers climb onto thick vertical perches such as tree trunks, while ant birds cling to slender vertical perches. Tanagers hunt from horizontal twigs and branches. Each species of bird has its preferred hunting spot. They don't often clash.

Ant-following birds served as great warning signals as I worked in the swamp forest near Belém, Brazil, years ago. On my first field day, I inadvertently stood in the path of an army ant swarm. Ants climbed up my boots, boiled beneath my jeans, and bit my bare legs. I ripped off my clothes and frantically brushed off the offenders as my Brazilian field assistants stared, openmouthed. Later when I shared my trauma with my boss, Tom Lovejoy, he advised me: "Pay attention to where you see ant-following birds, especially plain brown woodcreepers, white-shouldered fire-eyes, and black-spotted bare-eyes, because swarms of army ants could be nearby." It didn't always work (I still got attacked when I wasn't paying attention to where I stepped), but I'm sure my encounters with army ants were less frequent than if I hadn't kept an eye out for these common birds.

SOMETIMES TWO KINDS of animals hunt together for food, and both species benefit. Hornbills are medium to large Old World birds with huge, colorful bills like those of New World toucans. They eat insects, reptiles, and small mammals. Dwarf mongooses, which we met in the first section, eat primarily insects. When Anne Rasa first watched foraging dwarf mongooses in the Taru Desert of eastern Kenya, she saw both yellow-billed and Von der Decken's hornbills hop behind or in the midst of the mongoose group. As grasshoppers and moths flew up from the grass, the hornbills caught the insects in midair. Rasa at first assumed that the birds were social parasites, stealing prey from the mongooses. After studying this association for several months, however, she realized that both animals benefit.

The hornbills actively seek out the mongooses, which generally spend the night in ventilation shafts of active termite mounds (yet another interspecies relationship). In the morning, the birds wait in trees near the mound for the mongooses to finish grooming, defecating, and marking territories with their scent glands. Because mongooses rarely spend more than a single night in one termite mound, the birds presumably track their mongoose "partners" each evening. Rasa found that the hornbills became agitated when the mongooses prolonged their morning ritual . . . just like a husband anxious to leave for a dinner engagement while his wife performs her "going out to socialize" rituals. When the birds have already waited more than an hour for the mongooses to emerge from their mound, "a bird flies down onto the mound, peers into the ventilation shafts, and calls, making a rhythmic 'wok-wok-wok' sound. This approach gets a quick response; the mongooses appear immediately and leave the mound within four minutes." (Perhaps fidgeting husbands should try this.)

The mongooses also count on their morning rendezvous with the birds. If no hornbills appear, ready for the day's foraging bout, the mongooses take almost twice as long to leave their termite mound. One can imagine them sighing, "Well, crew, we're on our own today; but, hey, we still gotta eat, so let's just go without 'em."

Teaming up with mongooses provides the hornbills with easy access to insects and mice that the mongooses flush, while the mongooses enjoy an early warning of predators, especially birds of prey. Mongooses spend about 18 percent of their foraging time hiding from raptors cruising overhead. Many raptors that prey on mongooses also prey on hornbills.

If a raptor appears, hornbills "explode into vertical flight and land in the nearest tree, where they hide among the branches while making very slow 'wok' sounds." The mongooses run and hide. Dwarf mongooses also have their own guards that warn of danger, but the more birds in a foraging group, the fewer mongooses remain on guard, which means that more mongooses can forage more of the time.

Coordinated hunting is a relationship in which the two animals play different roles during the hunt. An example is the interaction between fishes of two species: giant moray eels and roving coral groupers. Giant moray eels can reach ten feet in length with a girth as thick as your thigh. They normally hunt alone at night by sneaking through crevices in the coral reef and cornering their victims. Roving coral groupers are robust-bodied brown to red fish with large mouths. These predatory fish, reaching nearly three feet in body length, normally hunt alone during the day in open water. In 2006 Redouan Bshary and three colleagues reported on coordinated hunting between these fishes in coral reefs in the Red Sea off the coast of Egypt.

The snorkeling biologists followed groupers and watched them approach giant moray eels ensconced in their crevices. The groupers "invited" the eels to hunt by shaking their heads several times per second within an inch or so from the morays. Sometimes groupers signaled eels before hunting excursions as if to say, "C'mon, let's get something to eat." Other times groupers offered the invitation after prey had escaped from them into crevices as if to say, "I've found something good. Come help me!" Fifty-eight percent of 120 eels obliged. And which got the food if the eel successfully rooted the prey out from the crevice? Sometimes the grouper, sometimes the eel. Either way, the lucky one swallowed the prey whole. The unsuccessful partner never acted aggressively. Overall, the two fishes were equally successful. Because the prey were swallowed whole and right away, there was nothing to fight over. If partners instead fought over the food, or if one species were more successful than the other, perhaps such behavior wouldn't exist.

An intriguing aspect of this interaction is that the groupers actively signaled the eels—a necessary step because the eels are normally inactive during the day when groupers hunt. Is this behavior learned or innate? The fact that eels were so variable in their response to groupers' hunting invitations suggests that the behavior may be learned.

Western Native American language, art, and folklore tell tales of "friendship" between badgers and coyotes. White Mountain Apaches in Arizona refer to the mammals as "cousins" that travel together, and

Navajos have long portrayed these carnivores as hunting partners. Early European explorers and pioneers, and later western cowboys and ranchers, wrote of the hunting partnership between badgers and coyotes. Few field biologists, however, had studied the interaction until Steven Minta and his colleagues frequently observed teams of coyotes and badgers hunting together for ground squirrels in Jackson Hole, Wyoming, during the 1980s.

Badgers are specialized for digging, and they have extremely good senses of smell and hearing. When a badger senses a ground squirrel under the ground, it digs down to the rodent's burrow and tries to corner it. These burrows have escape routes, however, so trapping its prey isn't always easy for the badger. In contrast, coyotes are opportunists, always on the lookout for prey. They capture rodents by pouncing and chasing—aboveground. When a badger and a coyote are both after the same ground squirrel, what's the rodent to do? With the coyote ready to pounce on the surface, the ground squirrel fears to come out—giving the badger a better chance to trap its prey. But if the ground squirrel does pop out, the coyote is ready and waiting. Minta and his colleagues found that in brushy habitat, where vegetation obstructed the coyotes' locomotion, many more coyotes hunted with badgers than hunted alone. Coyotes hunting with badgers caught about one-third more rodents than did lone coyotes. Because badgers with coyotes spent more time underground, presumably they spent less energy moving about and digging. A win-win situation for both parties.

WE HUMANS TRAIN other animals to hunt for or with us. Our most common hunting partner is the dog. Humans domesticated dogs more than 10,000 years ago and over the years have bred a variety of hounds and sporting dogs, capitalizing on dogs' intelligence, loyalty, and tracking ability.

Hounds hunt either by sight or by scent. Whippets and greyhounds were bred to hunt by sight and then chase down game by running full speed. Greyhounds were developed in Egypt between 4000 and 3500 B.C. to hunt gazelles. Bloodhounds, another old breed, were developed in the Middle East about 100 B.C. to hunt by scent. Bassett hounds also run with their noses to the ground and follow prey by scent (until they trip over their own ears . . .). They were bred in France for not-so-wealthy countrymen who hunted on foot and needed a

dog with a keen nose that wouldn't outrun them. Another scent-hunter is the long-bodied, short-legged dachshund. "Weiner dogs" were originally bred in Germany to be badger hounds (*dachs* is German for "badger"), but they were also used to slip into rabbit burrows and drive out the prey. As any dachshund owner knows, this breed is exceedingly stubborn—a trait that serves it well as a hunter. While underground, a dachshund doesn't give up but keeps digging and digging. (Conan, my long-haired dachshund, shows his stubbornness by refusing to respond to any command . . . unless there's food involved, when his hearing suddenly improves.) English foxhounds—bred for speed, leaping power, a loud voice, and willingness to hunt in a pack—track foxes and other prey by scent, followed by hunters on horseback.

Sporting dogs—such as pointers, setters, retrievers, and spaniels—assist bird hunters. Many of these breeds were developed in Europe during the 1800s. Cocker and English springer spaniels dive into bushes and flush out birds. They stand motionless while the hunter shoots, then retrieve the prey on command. Setters and pointers guide hunters to hidden birds by using their sharp eyesight and keen sense of smell, then freeze in position and point their bodies toward their finds. Labrador retrievers pick up birds or other game that have been shot, usually from the water, and bring them back to the hunter.

Portuguese water dogs were first bred in Portugal in the 700s, to work with fishermen. These webbed-footed dogs drive fish into nets, retrieve lines, and retrieve fish from the water. They're also good at carrying messages between boats and at rescuing fishermen who have fallen overboard.

We train more than dogs as hunting partners. Cormorants, medium to large birds with sharply hooked bills, are excellent swimmers. They dive from the water surface, propel themselves underwater by kicking their webbed feet, and then surface to swallow their catch. Cormorants are so skilled at catching fish that for over a millennium Chinese and Japanese fishermen have trained them as fishing partners.

Near the town of Wucheng in southeastern China, 22 families fish with about 125 cormorants. The fishermen often clip their birds' wings, and they tie a three-foot string around one leg to ensure the bird doesn't take off. Before releasing a bird to dive for fish, the fisherman ties a grass

straw around its neck to ensure that if the
cormorant swallows a fish before the fisherman can grab it, the fish won't pass all the
way down the throat. The fisherman
strokes the bird's neck upward, and
the bird gives up its catch. Usually
though, the fisherman can snatch the
fish from the cormorant's beak while
the bird is positioning its catch to swallow it headfirst.
On a good day, a cormorant fisherman can catch 100 pounds of fish.

In China most fishing cormorants are bred and raised in captivity.
Training begins at about six months of age, and by one year birds are ready
for serious fishing. The birds can live 25 years. Once the birds are too old
to be productive, some fishermen allow their birds to come along for the
ride and fish for themselves. When a bird's time finally comes, a fisherman from the Li River honors his hunting partner with a celebratory death
feast. The man serves the bird a pound of meat and a pound of fish. He
then euthanizes his fishing partner with a quart of 60-proof spirits and
buries it in a little wooden box.

In some hunting relationships between humans and other animals,
both partners benefit. In 1996 Brian Smith verified a tale he had read in
an 1879 monograph by the British naturalist John Anderson: Fishermen
in Myanmar (formerly Burma) claimed that dolphins purposely drew fish
into their nets. Smith hired fishermen U Than Htun and his son, Myint
Kyaw Oo, to join him for a week of fishing on a research boat. U Than
Htun told Smith that with the dolphins' help he and his son catch as
many fish in a single cast as they catch in an entire day on their own. The
dolphins gain also, as they easily catch fish confused by the net and fish
stuck on the muddy river bottom after the net is pulled up.

The father described how to summon the dolphins as follows:

*You tap the sides of the canoe with a labai kway (a cone-shaped wooden pin), and slap
the surface of the water with the flat end of the paddle. Then you dangle the lead weights
attached to the net against the bottom of the boat, while calling out with the guttural
voice of a gobbling turkey. If the dolphins agree to help, one of them will slap the water
with its tail flukes. Then, one or two animals will take the lead and swim in smaller and
smaller circles to corral the fish to swim toward shore. The dolphins deliver the fish with
a wave of their half-submerged flukes. That's the signal to throw the net.*

Dolphins don't always cooperate, though. During the week, the team
came across three groups of dolphins. The first two groups ignored the call

and swam away. The son's response: "Getting the dolphins to cooperate is like seducing a young girl. Sometimes your efforts pay off, but other times she leaves you waiting." (And this from a boy who had not yet reached puberty . . .) Just as the expedition was ending, a group of three dolphins appeared. Father and son invited the dolphins. One dolphin slapped the water with its tail, and another herded a school of fish toward the fishermen. Smith writes: "U Than Htun threw his net. The dolphins rushed to grab the fleeing fish. U Than Htun let them have their fill, then slowly pulled up the net. It looked like a Christmas tree decorated with silvery ornaments." Father and son broke into huge grins. They had proven to Smith that dolphins can team with humans to catch fish.

Dolphin and human fishing partnerships also occur in parts of the Mediterranean, Mauritania, and southern Brazil. John Downer describes a Brazilian interspecies partnership in his book *Weird Nature*. Fishermen wait on shore for a dolphin's cue to know when to throw their nets. When dolphins find a shoal of mullet, they signal with a lunging dive and then herd the fish toward shore. The fishermen get a windfall, but the dolphins also feast on mullet panicked by the nets. Downer writes: "This cooperative fishing began several hundred years ago, and young dolphins learn the technique by watching their parents. Likewise, the fishermen pass on their knowledge to their children, and every dolphin is given a name. In this way the techniques involved are handed down, in the families of both humans and dolphins, from one generation to the next."

Another example of hunting mutualism involves humans and honeyguides, members of the woodpecker family. These birds are widespread in tropical Africa, where they eat insects, spiders, and occasionally fruits. Some eat bees and their larvae. Certain honeyguides are among the few birds that eat wax—a peculiarity noted by a Portuguese missionary in the mid-1500s who complained that honeyguides ate the candles he set out on his church's altar.

In the 1700s indigenous Africans claimed that honeyguides call and then guide ratels, badger-like mammals also called "honey badgers," to bees' nests. This has become a "fact" entrenched in natural history lore. The ratel supposedly breaks open the hive with its long claws and eats the honey. Afterward, the honeyguide feasts on the wax and larvae from the broken honeycombs. No biologist, however, has ever observed this behavior, and many doubt that this close relationship exists. Bee colonies are very common, and ratels probably have no trouble finding them on their own.

What is a fact, however, is that greater honeyguides (appropriately given the scientific name *Indicator indicator*) guide humans to beehives. The earliest written accounts of this relationship also date back to the 1700s. Recently, Hussein Isack and a colleague studied greater honeyguides and the nomadic Boran honey-gatherer people of northern Kenya. The birds lead the Borans to bee colonies. After the people have opened the nest and gathered honey, the birds extract larvae and wax from honeycomb left behind—food that supplements their diet of insects. When hunting for honey in unfamiliar areas, it took Borans, on average, nearly nine hours to find a beehive when not guided by birds, versus a little over three hours when guided. Obviously, the Borans benefit from the honeyguides' help. The birds benefit as well. Without the help of people opening nests, only 4 percent of hives would otherwise be accessible to the birds. In addition, the smoky fire that the Borans use when opening hives reduces the birds' risk of getting stung.

To get the birds' attention, Borans give a penetrating whistle by blowing into clasped fists, modified snail shells, or hollowed-out palm nuts. The birds signal humans by flying close, moving restlessly in the bushes, and calling *"tirr-tirr-tirr-tirr."* People and birds also communicate frequently while traveling. Isack and colleague write: "If approached to within 5 to 15 m, the bird takes off, still calling. After a short undulating flight, during which the white outer tail feathers are displayed, it perches again and continues calling. As the Borans follow, they whistle, bang on wood, and talk loudly to the bird to keep it interested in the guiding. When they get close to it, the bird flies to another perch. This pattern of leading and following is repeated until the bee colony is reached."

Borans claim that honeyguides inform them about the direction to the bee colony, about the distance to the colony, and when they've reached their destination. The birds do this through different calls, flight patterns, where they perch waiting for the honey gatherers to catch up, and timing of when they appear and disappear. Isack and colleague tested the Borans' claims and found all to be true.

DON'T THINK THAT any of the animals described here (including the humans) would starve without the other. Any one is perfectly capable of finding a meal on its own. It simply eats more heartily in mixed company. For humans, canine hunting partners add sport and enjoyment. Many hunters train dogs not for the purpose of increasing the number of quail or pheasants they can bag, but for the sheer fun of sharing outdoor adventures with their companions.

TAKEN TO THE CLEANERS

Cleanliness is, indeed, next to godliness.

JOHN WESLEY, 1778 sermon

As a biologist, visiting the Galápagos Islands a few years ago was a thrill of a lifetime. My favorite animals were marine iguanas, the world's only seagoing lizards. I spent innumerable hours watching them from within touching distance. Most of the time they snoozed on the black volcanic rocks, but every once in a while one snorted salt water out its nostrils or dove into the water after a bite of seaweed. I soon decided that Charles Darwin's description of these lizards in *The Voyage of the Beagle* did them an injustice: "It is a hideous-looking creature, of a dirty black colour, stupid and sluggish in its movements." Depending on the island, these stocky lizards are coal-black, dark gray, or dark brown. During breeding season, males add splashes of green, tan, or blood-red. Up to five feet long, these prehistoric-looking reptiles have chunky heads, expressive dark eyes, and a row of large, spiny scales running atop the backbone from behind the head to the tip of the tail. Handsome creatures, in my opinion.

Although bloodsucking ticks plague marine iguanas, the lizards have de-ticking helpers: Sally Lightfoot crabs. These bright orange or scarlet crabs, named for their habit of quickly scampering over the water surface or rocks to escape danger, popped up everywhere I watched the iguanas. The crabs eat algae scraped from rocks in the intertidal zone, but they also have a fondness for ticks. As the salt-encrusted lizards dozed on their hot rocks in the sun, Sally Lightfoots scurried over the reptiles' tough, scaly hides. The crabs teased with their claws until they removed embedded parasites.

The great American naturalist William Beebe also was impressed with Sally Lightfoots eating ticks from marine iguanas. In his 1924 account, *Galápagos: World's End,* Beebe comments that if his companion had not also witnessed the behavior, he "should hesitate to record such a remarkable occurrence." Beebe writes: "The crab had reached the head of the iguana, and instead of turning aside, crawled straight ahead, the lizard closing its eyes to avoid the sharp legs of the crustacean. On and on the crab went, slowly descending the whole length of the lizard. Three times it stopped and picked a tick from the skin beneath it, the black tissue being pulled high as the crab tugged away."

OTHER ANIMALS REMOVE ectoparasites (parasites on the outside of the host's body), mucus, scales, or dead skin from another animal species. Certain crabs, shrimps, fishes, and birds perform this service for clients that range from octopuses to black rhinoceroses. In each case, the cleaner gets a meal and the client gets cleaner.

Shrimps from three different families clean ectoparasites and dead tissue from fishes and other hosts. Temperate-zone cleaner shrimps provide their service when the opportunity arises, but they don't advertise for clients. In contrast, specialized tropical cleaner shrimps have extraordinarily long antennae that they wave to attract clients. Some establish cleaning stations.

California cleaning shrimps wander about at night in groups of up to several hundred individuals, gleaning bits of food from the substrate. When they come upon a "dirty" lobster, they scurry over the animal's body and eat decaying tissue, ectoparasites, and anything else edible. They do the same for California moray eels that wander into their daytime shelters. The shrimps sometimes enter the eels' mouths, but not without risk. Often the eels eat them. This cleaner-host relationship still needs some adjustment.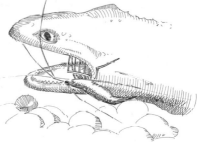

One specialized cleaner is the sedentary and solitary Pederson cleaner shrimp, which uses sea anemones as cleaning stations. This transparent 1½- to 2½-inch shrimp, decorated with violet spots and white stripes, perches among an anemone's tentacles or lives alongside an anemone in a rock crevice. Fishes quickly discover the shrimp's presence. They line up or crowd around, waiting their turns at the anemone station and staying put even when the cleaner shrimp takes a break. As a fish approaches it, the shrimp whips its long antennae and sways its body back and forth. If eager for a cleaning, the fish swims to within an inch or so of the shrimp. The shrimp climbs onto the fish and scuttles over its nearly motionless body, dislodging and eating attached parasites or dead skin from wounds. As the shrimp moves to the gill covers, the fish opens them one at a time as the shrimp enters and forages around the gills. The fish opens its mouth, and the shrimp enters and cleans.

Once its freshly cleaned client swims away, if still hungry the shrimp cleans the next fish in line.

Several species of tropical cleaner shrimps, including banded coral shrimp and Pacific cleaner shrimp, make popular saltwater aquarium pets—they keep both the aquarium and their aquarium-mates clean. You can get cleaned also. Stick your hand in, and they'll pick at your skin and even clean under your fingernails searching for food. One cleaner shrimp is a Hollywood star. If you saw the animated movie *Finding Nemo*, you'll remember Jacques, a Pacific cleaner shrimp.

SOME FISHES CLEAN other fishes. Over 110 species of small fishes, including gobies and wrasses, feed on ectoparasites and dead tissue from larger animals. Most cleaner fishes live in marine tropical waters. They have pointed snouts and teeth that work like tweezers to pluck off parasites. Their bright, conspicuous colors and patterns advertise their profession. Some tropical cleaners swim forward, turn sideways, and then retreat, repeating the display until a client settles in to be cleaned. Many set up cleaning stations where their clients congregate and wait to be serviced, as in the case of Pederson cleaner shrimps. Cleaners extend invitations, but clients also make requests. Clients often display or pose to attract the cleaners' attention and to indicate that they won't be eaten. Once cleaners and clients have found each other, the cleaners swim close, inspect, then nibble away.

Many clients normally eat small fishes, yet they open their mouths and allow the cleaners to swim in and go about their business unharmed. Some clients raise their gill covers and allow cleaner fishes to enter the chambers and clean the gills, just as the shrimps do. Wrasses, brightly colored cleaner fishes with distinct patterns of dark stripes, clean sharks, barracudas, and other large predatory fishes—even moray eels—that appear to be mesmerized by the cleaners' gentle nibbling.

Imagine snorkeling on a coral reef. You watch a three-inch Red Sea wrasse cleaning ectoparasites from the head of a four-foot moray eel, one the most voracious fishes known. The eel opens its mouth. The tiny wrasse enters and removes parasites. Once finished, the wrasse saunters out from the gaping mouth. You continue to watch as a small non-cleaner fish swims near the eel and is gobbled up. How does the eel distinguish between cleaners and snack food? Most likely through a combination of the cleaners' distinct color pattern and posture.

A friend of mine, Kay, had an intimate encounter with a wrasse while snorkeling with her son on Australia's Great Barrier Reef: "Shea and I were snorkeling when, about eight to ten feet below us, we saw a wrasse at its cleaning station. Several big parrotfish waited to be cleaned. Arms floating, fins motionless, we watched, enthralled, as the slender blue, black, and white-striped wrasse cleaned the big-beaked fish, searching their backs and gills. After the wrasse cleaned several fish, it looked up and swam straight to us. It inspected the seal of my face mask, checked out Shea's armpits, and returned to its station, where the parrotfish had waited patiently. Maybe the little fish wanted to see close up what strange, ungainly creatures were shading its station, but Shea and I were convinced it thought our turn had come to be serviced. We bragged all day that we were wrasse-clean."

Although cleaner gobies in the West Indies and Puerto Rico generally feed on parasites from fishes, there is a report of gobies cleaning a large octopus in the Caribbean off St. Croix. The observers watched two gobies pick repeatedly at the skin around an octopus's head and body. At least twice, one of the fish entered the octopus's siphon, the tube through which it expels water. This happened out in the open, during the day—the first time the observers had seen an octopus in the open by day. The octopus seemed perturbed by the observers' presence, but reluctant to seek cover. Personal hygiene came first.

In addition to parasite removal, cleaner fishes might help heal wounds. Coral reef fishes suffer frequent minor injuries. Sea urchin spines penetrate their muscles. Corals cut and abrade their scales. Yet infection rarely occurs. Susan Foster studied injury and infection in the Caribbean blue tang, a coral reef surgeonfish, on reefs off San Blas Island, Panama. During 23 observation days, she found that nearly 14 percent of 118 adults had minor cuts and scrapes, yet she never saw fungal or bacterial infection in these fish or any other individuals. Foster suggested that lack of infection might be due to cleaner fishes removing dead tissue.

During her eighteen-month study, Foster saw only three severely injured adult blue tangs. She followed the healing process of these individuals and found that during the first five days after injury each fish spent more time at cleaning stations than they did later, once black "scabs" had formed over their wounds. Three species of cleaners—bluehead wrass, Spanish hogfish, and a cleaner goby—took bites from the wound edges, bit at detached muscle strands and other damaged tissue, and occasionally took bites from the surface of the "unscabbed" wounds. All three tangs healed completely and without infection.

SOME BIRDS EAT ectoparasites from hosts. Egyptian plovers maintain a special relationship with carnivorous crocodiles. While the crocs sunbathe on riverbanks, the plovers pick off and eat parasites from the reptiles' scaly skin. Over 2,500 years ago the Greek historian Herodotus claimed that bloodsucking leeches attach themselves to the crocodiles' gums, and there's nothing the reptiles can do about it—except sun themselves, open their mouths, and wait for plovers to help them out. He wrote that the birds hop into the crocodiles' mouths, where they snap up leeches and other parasites. We now know that the birds are not specialized dental hygienists. They eat ectoparasites wherever they can find them on crocodiles.

Birds clean other reptiles as well. Certain finches on the Galápagos Islands eat ectoparasites from tortoises and marine iguanas. In the case of the Galápagos tortoises, the finch hops in front of its host, apparently to signal "I'm ready to clean you." In response, the tortoise raises high on its elephantine legs, lifts its shell off the ground, and slowly stretches out its neck. Then the finch can easily dig out and eat the ticks and other parasites from every nook and cranny of the reptile's wrinkled skin.

One of my favorite sights while canoeing on the River Styx in northern Florida was watching freshwater turtles bask. I related to these reptiles. How lovely to while away the afternoon, legs outstretched on a log, soaking up the sun's warmth and energy. But basking offers turtles more than pleasure and relaxation. By warming up, turtles can digest their food faster, and the sun's rays retard algal and fungal growth. Another plus might be that as turtles bask, their skin dries and attached bloodsucking leeches lose water and fall off. Richard Vogt suggested that, at least for map turtles, basking with outstretched legs might also give grackles, medium-sized birds related to blackbirds, the opportunity to feed on attached leeches.

While watching turtles on the Mississippi River in Wisconsin, Vogt saw a grackle inspect the "armpits" and "legpits" of map turtles sunbathing on a log. When the bird found a leech, it pulled off the parasite and ate it. One particular leech was a challenge. "The object was attached so well that the turtle was rocked back and forth on the log as the grackle attempted to pull the object from the turtle. During this process the turtle's legs remained motionless." The grackle foraged intermittently for leeches on basking turtles for about 2½ hours, and the turtles cooperated by holding their legs outstretched and still. Vogt returned five days later and again watched a grackle harvest leeches from basking map turtles.

Oxpeckers, also called "tick-birds," seek out hippopotamuses, rhinoceroses, wart hogs, zebras, elephants, giraffes, buffalo, impalas, and other large mammals from which they eat ticks and parasitic fly larvae as well as maggots and dead tissue from wounds. Oxpeckers grip their hosts' hides and crawl over the bodies with their needle-sharp, strongly curved claws. They peck, probe, and grasp prey with their sharp-edged bills. Oxpeckers spend nearly all their waking hours on their hosts where they not only feed but also mate. Their entire diet comes from what they find on their hosts.

Randall Breitwisch watched oxpeckers de-tick mammals in the Masai Mara Nature Reserve in southwestern Kenya. Imagine: Six yellow-billed oxpeckers ride on a zebra, and fourteen perch on a giraffe's backbone. A red-billed oxpecker cleans the face of a Cape buffalo, an animal renowned for its bad temper. Seven red-billed oxpeckers work on a black rhinoceros. Breitwisch describes the following interaction between a yellow-billed oxpecker and a zebra:

The moment the bird lands on the zebra's back, the animal abruptly stops its mechanical feeding, lifts its head high, and stands still and erect. The brown, starling-sized bird begins to act like a huge probing insect. Looking for all the world like a flea wending its way across a forest of hair, the oxpecker skitters across the zebra's body. The oxpecker moves from back to flank, dips under the belly (where it hangs upside down), works down one leg to the hoof and up to the side, and completes its circuit on the back, where it perches and preens in the morning sun.

Do oxpeckers really improve the personal hygiene of these large mammals? Breitwisch points out that to measure health benefits we would need to prevent oxpeckers from cleaning their hosts and then document the health of these uncleaned animals, an experiment that hasn't been done for various reasons (think rhinoceroses . . .). We know, however, that a single tick can drink enough blood from a domestic calf to decrease the animal's growth rate by more than a pound a year. One tick! Imagine the possible effect of 100 ticks on a baby zebra, giraffe, or even Cape buffalo.

Breitwisch suggests that the mammals not only tolerate the oxpeckers' cleaning activities; they also encourage them by making it easy for the birds to clean—even in the most intimate body regions. An oxpecker gives a stereotyped display of hopping along a zebra's back and then perching on its rump. The zebra responds: "Still in its statuelike pose, the zebra raised its tail quickly and held it out behind its body. The oxpecker immediately scuttled down onto the hairless, black shiny skin that surrounds the zebra's anal region and worked this area assiduously." Breitwisch watched oxpeckers hop down the back and perch on zebras' rumps 134 times. In over two-thirds of these observations, the zebra raised its tail. The only other time a zebra raises its tail is to defecate. Apparently, the two animals—oxpecker and zebra—communicate. Both gain from the grooming relationship.

Breitwisch watched various mammals appear almost mesmerized while being cleaned by oxpeckers, much like the behavior described for coral reef fishes being cleaned by wrasses. He writes: "I have seen zebra foals, impalas, and even tough little wart hogs apparently lose their balance and gently drop to the ground when being cleaned by oxpeckers. To all appearances, they have fallen under an avian spell."

MAYBE IF WE humans gave up our fast-paced lifestyle and instead roamed about slowly, over time we'd attract our own little avian cleaners. Just a random thought. But if even warthogs swoon from the experience and moray eels seem to be mesmerized, what might we ourselves experience, in addition to sparkling clean bodies?

SHE'S GOT A TICKET TO RIDE

Crocodile Song

She sailed away on a sunny summer's day,
On the back of a crocodile.
You see, said she, he's as tame as he can be,
I'll ride him down the Nile.

TRADITIONAL CAMPFIRE SONG

In the real world, some animals ride on other animals—a practice scientists call "phoresy"—to find food or mates, to disperse, or to locate egg-laying sites. Some hitchhikers pay nothing for their rides but do no harm. Other hitchhikers provide a service in return, and still others harm their hosts.

MY HUSBAND, PETE, gently untangled the green hermit hummingbird from the mist net. As he held it in his hand, about to record its weight, age, and size, three tiny nearly translucent mites suddenly erupted from the hummingbird's nostrils, ran to the tip of its long curved bill, then did an about-face and scurried back to their nostril haven. With a hand-made aspirator, Pete carefully sucked out the tiny passengers to give to his biologist friend Rob Colwell.

Hummingbird mites, eight-legged creatures related to ticks and spi-ders, are about twice the size of the period ending this sentence. These mites live, feed, mate, and lay eggs in hummingbird-pollinated flowers. But these flowers may last only a single day, at most a week, before falling off or shriveling up. The mite that doesn't relocate in time goes down with the flower. Sometimes the mite can simply switch to another, younger flower on the same inflorescence. Often, though, it catches a ride to another inflorescence or another plant some distance away. How? It dashes up the bill and into the nostril of a hummingbird that is momentarily drinking nectar from the home flower. Hitching a ride on the hummingbird "airbus" requires perfect timing, agility, and speed. Rob Colwell and his colleagues have es-timated that in terms of relative size, a mite must run as fast as a cheetah to get into the hummingbird's nostril before the bird withdraws its bill and zooms off—about twelve body lengths per second.

A mite stays in the bird's nostril, often jostling with fellow passenger mites, until it gets off at one or another airbus stop—as the humming-bird feeds at another flower, then another, then another. It must "decide" within a fraction of a second to five seconds whether to stay seated or disembark before its airbus takes off again. The split-second decision isn't simple. A hummingbird might visit several species of flowers within a few minutes, but the mite must jump off at a flower of its particular host plant species. The mite probably recognizes its preferred kind of flower by odor, "smelled" through tiny hairs on the tips of its forelegs. Colwell and his colleagues experimented with thousands of mites and found that no more than 1 in 200 disembarked at the "wrong" flower.

Experiments showed that each mite species preferred its own host nec-tar to that from other flower species. When forced to live on non-host nec-tar, though, many mites survived and reproduced just fine. Why, then, are these tiny hitchhikers so particular? The investigators suggested that mites seek not only food but also mates in host plants. Think in evolutionary time. Once a particular species of flower becomes the most popular host for a given mite species, individuals that jump off at other kinds of flowers won't often find mates and thus won't leave many offspring. But mites that jump off at popular hangouts—flowers of the host species—will encounter more mates and leave more descendants.

In another commensalistic drama, giant harlequin beetles rather than hummingbirds play the part of airplanes. Pseudoscorpions are the passengers. The setting is lowland tropical forest of Central and South America.

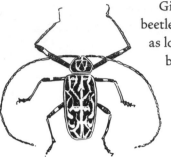 Giant harlequin beetles are a type of longhorned beetle, so named because the antennae are at least half as long as the body. Swirls, squiggles, and dashes of black, greenish yellow, and crimson or orange decorate harlequins' three-inch bodies. The passengers, pseudoscorpions, are tiny, flattened, oval-shaped arachnids with large claw-like pedipalps (the second pair of appendages). Most are less than one-fifth of an inch long. Like most pseudoscorpions, the ones that ride giant harlequin beetles live in decaying trees. To get from trees in the last stages of decay to freshly dead or dying trees, the pseudoscorpions hitch rides under the wings of harlequin beetles as the beetles emerge from pupal chambers in the rotting wood. The pseudoscorpions pinch the beetles' abdomens, which stimulates them to separate and open their wings. Once the hitchhikers climb aboard, the largest male pseudoscorpions shove the smaller males off before the flight begins. The beetles take off to seek out newly dead trees where they will mate and lay eggs.

While in flight, the beetles serve as battlegrounds and sexual playgrounds for their pseudoscorpion passengers. Male pseudoscorpions that succeed at staying on board compete with one another for females by establishing territories on their hosts. Then it's mating time until the beetle lands.

Upon landing, female pseudoscorpions quickly deplane. The males, though, may stay on board for two weeks or more and continue to defend their mobile sex nooks and mate with new females as they board. Males don't fall off during the numerous takeoffs, flights, and bumpy landings because they strap themselves to the beetles' abdomens with complex "safety harnesses" they construct from silk. Females attach themselves with simple seat belts made from single silken threads. Their flights are shorter and less risky.

What's the cost of a ticket? Nothing. The harlequin beetle provides shuttle service and a pleasure palace. The passengers ride for free but cause no harm.

AND NOW A MUTUALISTIC relationship. Imagine carrying a housecleaning service attached to your body wherever you go. Some native rodents in Central and South America do just that. Their housecleaners are wingless brown rove beetles, nearly half an inch long. These beetles grab clumps of their host's fur in their mandibles, often attaching at the base of the rodent's ears, nape of the neck, or elsewhere on the head. The rodent takes no notice, even when a beetle scurries across its eyes and whiskers. The beetles, sometimes as many as thirteen per host, ride at night when the rodents are active. At sunrise, after their hosts have bedded down in their nests, the beetles hop off and search through the nesting material for fleas, ticks, and large parasitic mites to eat. Later, as the rodents are about to leave on their nocturnal foraging bouts, the beetles reboard.

While on their hosts, the beetles neither hunt nor eat. Why, then, do they reboard and leave the nest and its uneaten food behind? Apparently the beetles use their hosts to travel to unexploited nests. Presumably a given rodent uses a variety of nests. So, it pays to stay with the host on the off chance that the next sunrise will bring a nest full of fleas, ticks, and mites. This travel arrangement benefits both parties. The beetles land in the middle of a smorgasbord without having to search for it, and the rodents are relieved of external parasites hanging around their homes, looking for blood meals.

AN UNUSUAL PARASITIC relationship involving phoresy between wasps and fire ants includes trickery on two levels. Female wasps of the family Eucharitidae lay their eggs in leaves, buds, and fruits of plants frequented by fire ants. After the eggs hatch, the extremely tiny (maximum 0.20 × 0.07 mm) wasp larvae latch onto foraging ants. Presumably the larvae are so tiny that the ants don't notice their presence. The ants unwittingly carry the wasp larvae to their nest. There, the larvae attack the ants' pupae and live as parasites until they themselves pupate and emerge as adult wasps.

Ants rely heavily on odor to recognize nestmates and identify invaders. Thus, to infiltrate an ant nest, a parasite must smell like the host. An ant's cuticle (outermost layer of its body) is coated with lipids, including some that absorb odors unique to its particular colony. The wasp larvae likely pick up these lipids and odors from contact with the ants, so they end up smelling just like the ants. The fire ants don't kill the wasp larvae because they can't distinguish between the wasps and their own brood. In fact, the ants carefully tend the wasp larvae and pupae as if they were their own. The fire ants have been doubly duped.

HOW A HITCHHIKING relationship works out for both animals involved (commensalism, mutualism, or parasitism) may depend on the circumstances. Consider remoras, elongate brown fishes that live in warm oceans. The eight species of remoras range in length 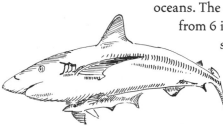 from 6 inches to 3½ feet. Remoras attach to sharks, rays, other large fishes, sea turtles, dolphins, and whales by means of large, oval suckers atop their heads. The suckers are fused to the upper jaws, forming part of the snouts— as if these fishes wore platters on their heads. Remoras attach to various sites, often to the host's belly or side, sometimes to the mouth or gills. They ride along on their hosts, traveling to food. These fishes leave their hosts for brief forays to feed on small fishes and invertebrates, and to mate and lay eggs.

Remoras gain transportation, but they also gain protection from predators just by associating with hosts that are much larger than they are. The hosts probably lose little by carrying remoras, except that it may cost them a little more energy to swim especially if transporting more than one remora. Depending on where the remora has attached, the hitchhiker may slightly reduce the host's hydrodynamic efficiency. Typically, neither of these costs is high enough for the hitchhiker to qualify as a parasite, however. But does the host gain anything? It depends on what the particular remoras eat. Some eat only scraps of food that fall from the host's mouth, bits of the host's vomit and feces, and free-living fishes and invertebrates. In this case, the host does not benefit, and the relationship is commensalism. Other species of remoras, however, eat ectoparasites, bacteria, and diseased tissues from their hosts, which benefits the host and tips the partnership toward mutualism.

Whalesuckers, remoras that hitchhike mainly on dolphins and whales, frequently attach to bellies, a position that presumably causes the least hydrodynamic drag. But sometimes they attach to other body sites and cause problems. Consider whalesuckers hitchhiking on spinner dolphins. Attachment close to a female's genital slit can prevent her from copulating. Whalesuckers attached below dolphins' eyes obstruct vision and may slow detection of shark predators. Remoras attached over part of a dolphin's blowhole may hinder respiration. In these cases, the remoras harm their hosts and thus act as parasites. But spinner dolphins aren't helpless. They can reposition inconveniently placed hitchhikers by leaping out of the water and spinning or doing tail-over-head flips. These maneuvers tend to relocate a whalesucker—ideally from an irritating place to a less sensitive spot. This is a great example of give-and-take between symbiotic species.

CHILDREN'S STORIES and folklore are filled with the magic of riding other animals, perhaps deriving inspiration from the natural world. Bellerophon rode the winged horse Pegasus to kill the Chimera, a fire-breathing creature with a lion's head, goat's body, and serpent's tail. One inch Thumbelina escaped the ugly mole and his gloomy underground passage by riding a swallow over the mountains to a country of permanent summer where lived a tiny, handsome prince. In *The Lion, the Witch and the Wardrobe*, Susan and Lucy rode the lion Aslan to Cair Paravel, Aslan's castle by the sea, where Aslan "de-stoned" a tall Dryad, unicorns, centaurs, dwarves, the Faun Mr. Tumnus, and Giant Rumblebuffin, among other creatures. Then they all fought the evil Snow Queen. Don't you agree that the true-life stories of animals riding on other animals to reach their goals are just as magical as the make-believe ones?

HOUSEGUESTS, UNLIKE DEAD FISH, DON'T ALWAYS SMELL IN THREE DAYS

If it were not for guests all houses would be graves.

KAHLIL GIBRAN

Gibran was referring to people, but other animals also welcome guests into their homes. As I contemplated which hospitality tales to include in this essay, a thought surfaced: make a burrow and someone else is likely to move in with you. Thus my focus on sharing burrows. Every time I see the opening to a burrow, I think, if not say aloud, "Who's down there?" Most of us know better than to stick our hands or fingers down a hole, but I admit I've poked sticks down burrows to entice the occupants out. It rarely works, though. Depending on the size and shape of the opening, the habitat and soil characteristics, and geographical location, the burrow-digger might be a worm, crab, wasp, beetle, mole cricket, spider, tortoise, groundhog, badger, or prairie dog, just to name a few. Chances are good he or she has guests.

In all of the relationships described here, one animal moves into the home made and occupied by another species, and the home owner doesn't object. It may even gain. The term "inquiline" refers to an animal that lives in the occupied dwelling place of an animal of another species. The word comes from the Latin *inquilinus,* meaning "an alien, one living in a place not his own."

BURROWS OFFER SAFE havens from many aboveground predators. They also offer quite stable temperature and humidity levels year-round, providing shelter during good times and bad in the outside world. One

extraordinary burrower is the pocket gopher. "Gopher" comes from the French *gaufre,* meaning "honeycomb," referring to the animal's tunnel system with many openings that lead to side tunnels used for pantries, bathrooms, and bedrooms. These little rodents, common throughout much of the United States, have fur-lined cheek pouches they use for carrying food—thus the common name. Cute, but . . .

Pocket gophers spend most of their lives underground in tunnel systems they dig with their large front claws and front teeth. Some complex tunnel systems extend 800 feet. These rodents eat plant roots and rhizomes (horizontal stems that grow at or just below ground level), and that explains why many people consider them to be extraordinary pests. Myself included. Within a year after my husband and I planted four young aspen trees, pocket gophers tunneled under them and chewed on the roots. Elk later finished off the weakened trees by eating all the leaves. Pocket gophers also ate the roots of and killed several young spruce and locust trees we planted. Pests they are, but I must admit it's captivating to watch an industrious pocket gopher throwing dirt out of its tunnel, making new excavations.

Certain other animals benefit from the burrowing activity of pocket gophers. At least 60 species of insects and other arthropods are inquilines in pocket gopher homes. Many have never been found anywhere else. One of the most common inquilines in some areas is the blind camel cricket, a pale insect with gangly legs that eats feces and organic debris. Blind camel crickets dig their own narrow passageways off the burrow walls. By staying in these, they can avoid getting buried as the pocket gophers push soil in their search for food.

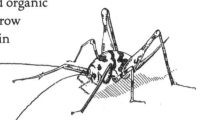

Gopher tortoises, from the southeastern United States, are other expert burrowers. Using their shovel-like front feet, these tortoises excavate burrows averaging fifteen feet long and six feet below the surface. At least 302 invertebrate and 60 vertebrate species have been found sheltering in tortoise burrows. Some are just chance drop-ins. Others are regulars, such as the gopher frog that depends on the burrows' moisture to survive. Another regular is the iridescent, blue-black eastern indigo snake, the longest snake in North America at a whopping record length of 8.6 feet. Imagine a snake this size sharing a burrow with a tortoise! Somehow they work things out. I like to think the snake's agreeable personality helps. My late friend and colleague Archie Carr, superb natural historian, was fond of indigo snakes. He described them as "handsome and extroverted" and suggested that "a more ingratiating house pet is hard to imagine." He and I (and our students) appreciated the fact that these snakes rarely bite when you pick them up. Agreeable personality, indeed.

In addition to a stable microclimate, gopher tortoise burrows offer their guests easily cornered food. When tortoises (not renowned as clean freaks) defecate in their burrows, their feces attract scat-eating invertebrates, which in turn provide food for houseguests such as frogs, lizards, and mice, which then provide gourmet dinners for the resident indigo snakes. Being vegetarians, the tortoises refrain from consuming their guests. The perfect hosts.

My favorite terrestrial-burrower story involves tarantula homes that attract strange bedfellows, an interaction studied by R. Howard Hunt. He vividly describes a snake/narrow-mouthed toad/tarantula interaction as follows:

Sensing another meal, the ribbon snake followed the toad's scent trail, which led to the entrance of the burrow. And just beyond, at the edge of darkness, crouched the toad. With lidless eyes fixed on its prey, the snake tensed for a lunge. Suddenly a great hairy body, propelled by eight hairy legs, burst from the depths of the burrow, and a fully mature, female tarantula scuttled to within an inch of the toad. Taking advantage of the situation, the toad fled to safety—beneath the imposing black fangs of the tarantula. The confused snake reversed itself and disappeared into the waving grass.

In Texas, Hunt once found a two-inch tarantula under a plate-sized fossil ammonite, along with six adult narrow-mouthed toads. He named the tarantula Ammonita and took her, the toads, and the ammonite to his laboratory, where he set them up in a terrarium. "Ammonita immediately started a burrow under the fossil. Working several nights like a mini-bulldozer, scooping out and carrying clods of earth with her versatile fangs, she excavated a spacious burrow six inches long and lined the entrance with silk." The narrow-mouthed toads moved into Ammonita's burrow, and the septet soon developed a routine. "In the evenings, the tarantula waited motionless at the mouth of the burrow, ready to snatch a passing insect, and one by one the six toads brushed by her and skulked around the periphery of the fossil."

Hunt released a ribbon snake into the terrarium.

As usual, the tarantula stood at the entrance of her burrow, and the toads foraged nearby. The snake soon neared the burrow, but the toads hopped toward their protector and the tarantula held her ground. At the moment of truth, only one toad remained under Ammonita, the rest having fled to the innermost recesses of the burrow. With one back foot resting lightly on the head of the toad, Ammonita slowly raised her fangs. The snake, evidently sensing no danger in this, crawled closer to the entrance, then tried to inch past Ammonita. Suddenly, the tarantula jabbed with feet and fangs. Startled, albeit uninjured, the snake sped out of the burrow and away from the thing lurking within.

The toad-tarantula relationship may be mutually beneficial. Narrow-mouthed toads gain food and protection from predators. The tarantulas may gain because the amphibians eat ants that otherwise might consume the tarantulas' eggs and hatchlings.

MARINE INTERTIDAL ANIMALS gain the same benefits from burrows as do terrestrial animals—shelter in a more stable environment and protection from many predators. Anyone who has spent time at the beach has seen the intertidal zone dotted with burrows. In Rachel Carson's words, "Sand . . . forms a yielding, shifting substratum of unstable nature, its particles incessantly stirred by the waves, so that few living things can establish or hold a place at its surface or even in its upper layers.

All have gone below, and in burrows, tubes, and underground chambers the hidden life of the sands is lived."

Larvae of parchment tube worms, six- to fifteen-inch segmented worms, build U-shaped tubes in shallow water along the coasts of North America and northern Europe. You may have seen these tubes on the beach, as storms often wash them out of the ocean. The two ends of the tube project several inches above the surface of the mud or sand. As the worm grows, it enlarges the tunnel, which eventually might reach 1½ inches in diameter and nearly 2½ feet in length. By fanning its body, the worm draws in water through one end of the tube and expels water from the other. This constant water current flowing through its home brings in oxygen and plant cells for food and flushes away feces.

Several commensal animals move in with these worms, the most common being two species of pea crabs (same size and shape as peas). It's hard to find a tube without a mated pair or at least one crab. Inside the tube, the crabs are relatively safe from predators and drastic fluctuations in the physical environment. They also are spared food shortages because they eat the minute food particles swooshing by in the worm-fanned current. Indeed, a pea crab may be down the tube for life. Pea crabs enter the burrows when young and eventually grow too big to leave the narrow exits. The crabs don't pay rent, but neither do they harm the worms.

Innkeeper worms are sausage-shaped unsegmented invertebrates that burrow. At least seventeen species of innkeeper worms share their burrows with houseguests. Most guests benefit from the food- and oxygen-rich water the innkeeper worms pump through their tunnels. Unlike pea crabs, some do good things for their hosts in return. For example, guest clams chow down on the impressive quantities of feces that innkeeper worms produce, thus keeping the burrows clean.

One of the best-studied innkeeper worms is the "fat innkeeper." This cigar-shaped worm burrows in mud at or below low-tide level in California mudflats. It digs with its short proboscis and scrapes the burrow walls with bristles surrounding its mouth. Then it crawls backward, using the bristles surrounding its anus for traction and blows the debris out the opening. Fat innkeepers generally grow to about 8 inches, though 20-inchers have been recorded. They live for up to 25 years, providing stable and long-lasting homes with plenty of food particles carried in by water currents they set up. Fat innkeeper burrows often house an impressive gathering of guests, including scale worms (flattish worms about half the length of your little finger), pea crabs, arrow gobies (small, bottom-dwelling fish), and hooded shrimps. All lodgers appear to be commensals, freeloading on the inn but not harming their innkeeper hosts.

Over the past several decades, biologists have learned much about the obligatory mutualistic relationship between at least 125 species of gobies and more than 30 species of snapping shrimps that construct and maintain burrows in the sand. During the 1970s, Lynn Preston studied the mutualism between a goby and two species of snapping shrimps off Hawaii. There, the shrimps live year-round in burrows as pairs, or sometimes trios of two females and one male. Gobies use these burrows as shelter. The goby gets a home, and the shrimps get an advance warning system of danger thanks to the goby's superb vision. Here's how it works.

The goby hangs out at the burrow entrance. When plowing out sand—burrows require continual maintenance—the shrimp leaves its burrow antennae first. It touches the goby's tail with its antennae, continues out of the burrow, and dumps the sand, then scuttles backward into the burrow. Preston interpreted the shrimp's touching the goby with its antennae as informing the fish of its presence outside the burrow—in effect saying, "I'm out and about. Please warn me of danger." If the goby senses danger while the shrimp is out of the burrow, it flicks its tail. That appears to be a signal, for the shrimp either freezes or flees into the burrow. A goby only flicks its tail when its partner shrimp is out of the burrow. Depending on the nature of the threat, the goby either stands its ground at the burrow entrance or it flees into the burrow—always *after* the shrimp.

In 1990 Yasunobu Yanagisawa reported on his observations of gobies and snapping shrimps off the island of Shikoku, on the southern coast of Japan. There, small burrows with openings a few inches in diameter are scattered along the sandy seafloor. A small, iridescent goby rests nearly motionless at the entrance to each burrow. Periodically, a small striped snapping shrimp emerges from a burrow and dumps grains of sand at the burrow opening. The shrimp spends more than 80 percent of its time making home improvements to the inside of its large branched burrow that may spread over an area of several square feet—quite a feat considering the shrimp is less than two inches long. The bond between fish and snapping shrimp is strong. The goby provides the shrimp with a tactile warning system of approaching danger, and the shrimp provides the fish with a burrow, a refuge during the daytime and a resting place at night. If isolated from its partner, neither fish nor shrimp will survive. Yanagisawa describes the shrimp's physical dependence on the fish as follows:

With its limited eyesight (its carapace covers its eyes), the timid snapping shrimp depends completely on its fish partner for protection outside the burrow. The shrimp maintains constant antennal contact with the goby, which it uses as a blind person uses a dog. When behind the goby, the shrimp touches the fish's tail fin with at least one antenna. When next to the goby, the shrimp bends an antenna sideways to touch the fish's pectoral or dorsal fin. When farther out of the burrow than the goby, the shrimp points one antenna backward to touch the fish's pectoral fin. If the goby swims too far away from the burrow, the shrimp gives up going out altogether and spends its time unloading sediment just within the burrow entrance, resuming its outside work only after the fish returns and contact is reestablished.

Yanagisawa provides observations and speculation on how the gobies and shrimps get together. Resident gobies ensconced in burrows don't tolerate baby goby intruders, so each baby goby must find a juvenile shrimp also starting off life on the sea bottom. Fortunately for them, both baby fish and shrimps appear on the scene from late July to late September. Yanagisawa writes: "When I crept carefully over the sea bottom during this period, I found tiny gobies, barely half an inch long, at burrow entrances only about one-eighth to one-fifth of an inch in diameter. Such small fish and burrows might be overlooked in an ordinary census. I also found shrimps, still semitransparent and smaller than the gobies, emerging from the burrows with loads of sand. When approached, the gobies and shrimps retreated rapidly into the burrows. They had already developed their tactile warning system."

Yanagisawa also found a few small juvenile shrimps alone near their fully excavated burrows, and a few apparently burrow-less baby gobies wandering about. These observations suggested to him that baby shrimps dig their burrows as soon as they settle on the bottom, and that juvenile gobies search about for available burrows. Individuals unable to form partnerships die. But those that get together still have a problem: each must find a mate.

The gobies reach maturity at nearly a year old. At that point, adults often move to nearby burrows to form pairs. The pairs stay together for several days before laying and fertilizing eggs. Males stay on for another several days to care for the eggs, but then the pairs split and search for other mates. And the snapping shrimps? Yanagisawa found that by two or three months, more than half the young shrimps he observed had already formed pairs. But how? Since the shrimps never venture outside their burrows without their goby partners, they could never go far enough during the day to find mates. And at night, they seal themselves into their burrows with sand. Yanagisawa suspected that the shrimps meet each other under the sand. As they enlarge their burrows, two neighboring burrows may join together. If the burrow owners are opposite sexes, they're all set. In Yanagisawa's words, "I believe that the shrimps' diligence in burrowing and their disproportionately large burrows are the result of their ardor to meet mates by means of this underground connection of adjacent burrows."

Ghost shrimps, two- to three-inch mud-dwelling whitish-yellow shrimps, also spend most of their lives in burrows they dig. Scale worms, pea crabs, gobies, and other animals share these homes. The California ghost shrimp hosts pairs of blind gobies. These fish live nowhere else and are completely dependent on their host shrimps. The fish stay in their hosts' burrows, eating bits of seaweed and animal matter. The female goby lays up to 15,000 eggs, glued to the burrow surface. Hatchling gobies have functional eyes, and they swim away from home. Six months later, membranes have grown over their eyes, and pairs slip inside ghost shrimp burrows to live out their lives together.

SHARING BURROWS IS just one example of animals using other animals' homes, but it's widespread. If gopher tortoises or innkeeper worms became extinct, would their inquilines seek other animals' homes? Might they strike out independently and, if able, make their own burrows? Or would they also go extinct? In large part, the answer depends on the nature of the relationship: convenient or obligatory.

In the following essay, we'll look at some animals that borrow bodies rather than burrows.

BE IT EVER SO HUMBLE

Mid pleasures and palaces though we may roam,
Be it ever so humble, there's no place like home.

JOHN H. PAYNE, "Home Sweet Home"

Homes. Some people prefer a simple log cabin. Others a Victorian gingerbread or a modern "trophy home." Regardless of our preference, homes provide shelter and protection. A sense of place. Stability. Security.

Some non-human animals never set up housekeeping. Others build nests, dig burrows, or hunker down in crevices, caves, or dens. And then there are the animals that call "home" the bodies of other animals. In some cases, these homes provide not only shelter and protection but also food. Consider the following examples of parasitism, commensalism, and mutualism in animals that use other animals' bodies as "home sweet home."

BOTFLIES, HAIRY PARASITIC FLIES about the size of honeybees, lay their eggs on horses, cattle, sheep, squirrels, and other animals. The larvae live in boil-like swellings on their hosts where they feed on tissue and fluids before dropping to the ground to complete development. In the tropics, some species of botflies infest humans and other primates. These botflies have an unusual way of getting eggs to their hosts. A female botfly catches a mosquito and attaches eggs to the mosquito's abdomen, then releases it. When the egg-carrying mosquito lands on a monkey, human, or other host to feed, the skin's warmth causes the eggs to hatch quickly into tiny larvae called "bots." The bots crawl down the mosquito's legs and chew pinhead-sized holes in the host's skin. They burrow inside, feed on tissues and fluids, and extend snorkel-like spiracles through the openings to breathe. The bots grow beneath the skin and eventually emerge, fall to the ground, and pupate.

I once served as a bot home in Costa Rica. After feeling a twinge of pain on my wrist, I looked down and saw the head of a grayish-white maggot bobbing up and down through a small opening. The bot had bumped up against, or nibbled on, a nerve. Had I yanked it out with tweezers, the bot would have hung on for dear life with its tiny but strong hooks. If its body had snapped, leaving some portion inside, I could have ended up with an infected wrist. Instead, I covered the breathing hole with masking tape. When I removed the tape a few hours later, the bot was halfway out, gasping for air. At that point I gently extracted the rest of its body with tweezers. The ⅓-inch bot now floats in a glass vial of alcohol.

Flesh flies are another group of parasitic flies that are particularly abundant in the tropics. They resemble large houseflies, except that they're hairy. Unlike botflies, the female flesh fly deposits tiny larvae directly on her host. The larvae burrow into their food source and chow down. One November day in Costa Rica, my graduate student and I found a dead harlequin frog sprawled on a mossy boulder along a stream bank. The frog had a maggot-infested ¼-inch hole in its hind leg. By the end of the day, we found five more parasitized, but live, harlequin frogs. Having no idea what these parasites were, we put the moribund frogs in plastic shoeboxes lined with several layers of wet paper towel. A week later the maggots had consumed most of the frogs' flesh and internal organs. All frogs had died. The maggots had left their hosts and had settled between the layers of paper towel. Soon the maggots pupated, and hairy black flesh flies emerged 17 to 30 days later.

Let's move on, from flies to wasps. Female parasitic wasps have needle-like ovipositors (egg-laying organs) protruding from their rear ends. They jab these into caterpillars and other insects and squeeze out their flexible eggs. The hosts then serve as shelter and food for the wasps' young.

One small shiny black parasitic wasp with a bright yellow abdomen does us a great favor: it parasitizes alfalfa weevil larvae. These larvae may look innocent—attractive green grubs with yellow-green stripes down

their backs—but they are actually ravenous vegetarians that consume and destroy alfalfa crops. The female wasp uses her saber-like ovipositor to inject a single egg into a weevil larva. After hatching, the wasp larva feasts on its host's fluids and tissues but leaves the vital organs alone. Meanwhile the weevil grub continues gorging on alfalfa for three or four weeks, growing to half an inch long before spinning its cocoon. At that point the wasp larva eats the remaining tissues, killing the weevil. The wasp larva spins its own cocoon inside the weevil's cocoon. Eating vital organs last clearly benefits the wasp, which thereby has time to grow and mature before its host dies.

Instead of laying eggs inside a free-ranging host, some parasites drag an animal to their nest before laying an egg in it. Two-inch black-and-yellow female wasps called "cicada killers" dig underground burrows with about sixteen brood chambers. They then search for cicadas, which they sting and paralyze. A wasp drags a cicada into each brood chamber and then lays an egg in the insect. When the eggs hatch a few days later, the larvae begin to devour their live but paralyzed hosts from the inside out.

What follows is a most amazing natural history—a host's nightmare. Or a Hollywood scriptwriter's dream! *Copidosomopsis tanytmemus*, another parasitic wasp, injects its eggs into caterpillars of the Mediterranean flour moth. But instead of each egg hatching into one larva, each egg produces 200 larvae! Although genetically identical, the larvae develop into two distinct castes. Some become highly mobile soldiers, equipped with strong jaws but without reproductive organs. These soldier larvae wander through the caterpillar host and attack and kill other species of parasitic wasp larvae. Once they have cleared the field, they die. Their slower-developing siblings, which are anatomically complete, consume the moth caterpillar and eventually emerge as adult wasps.

And now another amazing, bizarre parasite story hot off the press as I write—published in April 2008. This one involves nematodes (slender worms called "roundworms"), ants, and fruit-eating birds. The story starts when, several years ago while working in Panama, Steve Yanoviak and his colleagues found a colony of the arboreal ant *Cephalotes atratus* that included several foraging worker ants with conspicuous bright red gasters (the large, bulbous section at the hind end of an ant). All the other workers had black gasters. The investigators dissected the red gasters and found hundreds of transparent eggs in each one. Each egg housed a small, coiled nematode, which was later determined to be a new genus.

Later Yanoviak and his colleagues carried out detailed field observations and experiments. They found that parasitized ants nearly constantly hold their red gasters in a conspicuous, elevated position, and they have an erect and unstable gait. Parasitized ants are sluggish and release no detectable alarm pheromones when disturbed. Furthermore, the infected gasters are easily detached from the rest of their bodies. Infected workers are 10 percent smaller but 40 percent heavier than their uninfected comrades due to their parasite load, and the parasite causes weakening of the junction between the gaster and the rest of the body. Because the ants cling to twigs, a forager can pluck the gaster without removing the rest of the ant from the twig. The investigators offered free-ranging birds pairs of tethered ants: parasitized ants with red gasters and healthy ants with black gasters. Birds took many more of the red gasters, which resemble berries. When Yanoviak and his colleagues offered a chicken an infected ant, they found hundreds of intact nematode eggs in the feces between 2½ and 13 hours later. Furthermore, these ants eat bird feces. The investigators found that bird feces represented 68 percent of a sample of more than 300 food items carried by worker ants returning to their nest of an infected colony.

Based on these observations and experimental results, Yanoviak and his colleagues hypothesized the following scenario, possibly a unique nematode life cycle: A fruit-eating bird mistakenly eats what it assumes is a berry—a red gaster, made conspicuous by a parasitized ant's behavior.

The bird later deposits feces full of nematode eggs on a branch. A foraging *C. atratus* ant collects the feces and takes it back to its nest, where workers inside the nest feed it to some of the larvae. The nematode eggs hatch and the young develop within the ant larvae. Later, adult nematodes migrate to the gaster of ant pupae where they mate, thus completing the nematode's life cycle.

NOT ALL HOST-AS-HOUSE relationships are so rough on the host. Some pearlfish, so-called because their long, thin bodies have a pearly luster, live inside other marine animals. Sea cucumbers—invertebrates shaped like their namesake vegetables but with the consistency of soft leather pouches filled with Jell-O—make especially good homes because their elongate bodies mirror the pearlfish physique.

Pearlfish locate their sea cucumber hosts by sight. Then, getting up close and personal, the fish enter and exit through the back door, the host's anus, which they locate by smell. Some go into the anus headfirst, others tail-first. A sea cucumber breathes by drawing in water through its anus. The water passes through the animal's gills and then flows out the anus again. A pearlfish enters with the inflow. If the pearlfish gets the back door slammed in its face—or tail—as it tries to enter, the fish simply waits. The sea cucumber must breathe, so its anus soon reopens.

One well-studied species of pearlfish found in the Bahamas hides by day inside a species of six- to twelve-inch sea cucumber. The fish leave their hosts at night to feed on small shrimps and other invertebrates. While most fish return to their same host, some move around and enter different sea cucumbers. The fish seem to avoid hosts that already have tenants. For good reason. Pearlfish turn cannibalistic and eat their roommates under crowded conditions, a behavior usually not tolerated in human homes or college dorms.

And from the sea cucumbers' perspective? A pearlfish or two that enter and exit through the back door cause no hardship. But some sea cucumbers get stuck with certain species of pearlfish that live permanently inside their body cavities and nibble on their reproductive organs. Fortunately, sea cucumbers regenerate quickly. The nibbled gonads may patch themselves up and function just fine. Still, because patching and repair take energy, the relationship is parasitic.

SOMETIMES THE NATURE of the host-as-house relationship is so complex that the label we give it changes as we accumulate more and more understanding through careful study. The following is a splendid illustration of this. The interaction involves several species of minnows called "bitterlings"—found mainly in eastern Asia—and freshwater mussels. Early in the breeding season, a male bitterling selects a mussel and defends it against other males. Meanwhile, female bitterlings develop long tubular ovipositors. After a male and female pair up, the female places the tip of her ovipositor into the mussel's gill chambers and lays her eggs inside. Her mate then fertilizes the eggs by squirting sperm inside the mussel. After they hatch, the baby bitterlings live inside their mussel nursery for three to six weeks. The eggs and baby bitterlings have a safe place to develop and a constant supply of oxygen, but they have provided no benefit to the mussel . . . or so we thought. Based on these observations, naturalists originally described the relationship as commensalism.

Later it was discovered that the mussels use the bitterlings also. When the host mussel is a female, her clam shaped larvae attach themselves to the baby bitterlings just before the fish leave the nursery. The fish swim away carrying their tiny passengers. Eventually the mussel larvae drop off, scattering in all directions from their parent. The baby bitterlings "paid their rent" by helping disperse the mussel larvae, making the relationship mutualistic.

But the story gets more complicated. The mussel larvae aren't just getting a free ride. They also suck blood from the baby bitterlings. It turns out that these freshwater mussels normally expel their young directly into the water. When these free-ranging larvae bump up against a fish's gills or fins, their tiny hinged valves snap shut. The fish's tissue encysts the mussel larvae, which feed on their host's blood for several weeks. The mussel larvae die unless they attach to fish for this short period of time. How serendipitous for mussel larvae to share their nursery with bitterlings! Instead of being tossed out on their own to bump into fish by chance, they attach to their foster siblings. Knowing this, biologists termed the bitterling-mussel relationship parasitic.

The most recent twist to the story is a new bitterling-mussel relationship that is parasitism in the opposite direction—for now. In 2006 researchers reported on experiments with European mussels and a kind of bitterling that has recently invaded Europe from Asia. Exposing "invader" bitterlings and European mussels to each other, the investigators found that the mussel larvae rarely attached to the baby fish. Those that attached did not encyst. In the real world outside the lab, they would have died. The adult mussels provided a nursery but gained nothing in return. In fact, female mussels burdened with "invader" bitterling boarders grew more slowly. Smaller mussels produce fewer eggs. Thus, "invader" bitterlings are parasites on European mussels. Next experimental question: What happens when "invader" bitterlings are put together with Asian mussels? Answer: The mussels fight back. When "invader" bitterlings lay eggs in them, the Asian mussels rapidly snap their valves shut and expel a stream of water, tossing out babies with the bathwater.

Perhaps someday European mussels will evolve defenses against the bitterling invaders. And maybe bitterling species in Asia will evolve a way to deal with their baby mussel bloodsucking parasites. Not only does our understanding of host-as-house relationships change through time; the relationships themselves also change as parasites and hosts evolve.

FINALLY, LET'S TURN to mutualism in the host-as-house stories, and not only to mutualism but to romance as well. A sponge, the Venus's flower basket, is the host. This elegant, tube-shaped, foot-long sponge lives anchored to the ocean floor at depths of 3,000 to 5,000 feet in the South Pacific. Venus's flower basket is a "glass sponge," so-called because its skeleton is composed of needle-like spicules of silica. The sponge's body is a slightly curved tube that resembles a ram's horn made of delicately spun glass. The tenants are small shrimps that crawl inside when the sponges are young. Shrimps move in and out of sponges until they find a glass home with a roommate of the opposite sex. Once the shrimps grow to a certain size, they can't escape through the sponge's sides. And they can't escape through the tube's top because the sponge seals off this upper end as it grows.

The shrimps live and breed inside the sponge. Their tiny young leave home, seek young Venus's flower baskets, and begin the cycle anew. The shrimps gain a safe place to live and raise a family, and they eat food that circulates within the sponge. In return, the shrimps serve as full-time, live-in maids, keeping the sponge clean of bacteria and debris.

In various Asian cultures, the Venus's flower basket and its shrimps symbolize eternal love. In the Philippines and Japan, live Venus's flower baskets and their occupants are given as wedding gifts. Never mind that the shrimps are locked in their prison and have no alternative but to be faithful. They stay mated forever, symbolizing eternal married happiness. This is one relationship I hope withstands further scientific scrutiny and remains one of mutualism.

FOR ANIMALS THAT live inside frogs, caterpillars, ants, sea cucumbers, mussels, and Venus's flower baskets, there's literally "no place like home." These animals would probably die if evicted. But no relationship is risk-free, and living homes don't last forever. If the host gets sick, the tenant might not fare so well either. And if the living home gets eaten, the tenant goes down the hatch also. Dependency has obvious drawbacks. As the English poet and dramatist John Gay advised, "There is no dependence that can be sure but a dependence upon one's self."

RAISING THE DEVIL'S SPAWN

This [brood parasitism in cowbirds] is a mystery to me; nevertheless, my belief in the wisdom of Nature is not staggered by it.

JOHN JAMES AUDUBON, AS QUOTED
IN JANET LEMBKE, *Despicable Species*

Audubon was only one of many naturalists to be baffled by cowbirds dumping their eggs in other birds' nests and then slinking away, leaving the foster parents to do all the work. In her book *Despicable Species*, Janet Lembke writes that what cowbirds do is the "birdly equivalent" of leaving an unwanted human infant on the doorstep of a church or hospital, but she points out two significant differences: "First, the human mother practices parasitism on her own kind, while the brown-headed cowbird ventures forth and is parasitic only on species other than its own. Second, and more important, the human action is facultative, while that of the bird is obligate. In other words, the woman has options, but, in the wisdom of nature, the bird has been given no choice at all."

Even though they have no other options, cowbirds are maligned by people who call them "shifty birds" or "lazy birds" and consider them to be avian social outcasts. The birds are seen as cheaters, and most of us don't cheer for cheaters. Still, the birds' lifestyle is an intriguing one and one that serves them well, judging by their expanding range.

The behavior of laying eggs in another animal's nest and leaving the parenting to the host is called "brood parasitism," a form of social parasitism. Parental deadbeats include at least 1 percent of bird species, certain insects, and some freshwater fishes, among others. Let's begin by taking a closer look at brown-headed cowbirds—those "shifty birds."

FEMALE BROWN-HEADED COWBIRDS, ranging from Canada to northern Mexico, average just 41 seconds to lay a brown-speckled egg in the nest of another bird species. Some female cowbirds puncture or remove one or more of the hosts' eggs before leaving their own eggs. Appropriately the genus of these scoundrels, *Molothrus*, means "intruder" in Latin. Over 200 species serve as unwitting hosts, including flycatchers, gnatcatchers, wrens, bluebirds, thrushes, thrashers, vireos, warblers, tanagers, cardinals, sparrows, meadowlarks, grackles, orioles, and finches. The host parents incubate the intruder's egg and feed the ravenous chick that hatches from it. Generally the cowbird chick and the host's brood share the nest. The cowbird chick's enthusiastic begging gets it the lion's share of the food. It grows rapidly, leaving the nest after 10 to 11 days, but its foster parents continue to feed it for another 6 to 18 days. Unburdened by parenting, a female cowbird can lay 50 or more eggs in a single breeding season.

One puzzle is how a young cowbird knows that it's a cowbird given that it was raised by non-cowbird parents and surrounded by their nestlings for up to four weeks after hatching. Perhaps in some other birds this would lead to an identity crisis, but young cowbirds don't imprint on the host species in the way that ducks, geese, and chickens imprint on the first living creatures they see. Experiments with female brown-headed cowbirds raised in captivity, isolated from the sight and sound of other cowbirds, have shown that species recognition is innate for them. Pure instinct.

Of the 130 species of cuckoos worldwide, nearly half are brood parasites. They're even more destructive of their hosts' broods than are cowbirds. Since at least the time of Aristotle, more than 2,300 years ago, naturalists have observed cuckoos leaving their eggs in other birds' nests. The female cuckoo flies to a songbird's nest, peers in, grabs an egg in her beak, devours it, lays one of her own eggs, and flies away—all within ten seconds. Some hosts abandon the invaded nests. Others feed and care for a ravenous foster chick, but raise no young of their own during that breeding season.

Michael Brooke and Nicholas Davies studied the common European cuckoo in Britain. This twelve-inch bird generally parasitizes the nests of smaller songbirds such

as reed warblers, meadow pipits, and European redstarts. Cuckoos lay relatively small eggs, but still, they are always larger than their hosts' eggs. The color of common European cuckoo eggs varies to mimic the eggs of the most common hosts in the area where they live.

A common European cuckoo egg hatches in about eleven days, usually before any of the hosts' eggs. The baby cuckoo instinctively evicts its nest-mates. Brooke writes, "The hatchling maneuvers the host's eggs—or if any have hatched, the young—into a hollow in its back, shuffles backward and up the side of the nest, and tips its cargo over the edge." The host parents may witness the foul deed, but the poor suckers still incubate and feed the cuckoo chick until it fledges, at which time it might well outweigh them. Brooke and Davies wondered: Do the hosts do anything that helps them to avoid being parasitized? Do the cuckoos try to circumvent their hosts' defenses?

Brooke and Davies played mother cuckoo. They painted cuckoo-sized plastic eggs with acrylics to mimic brown eggs of meadow pipits, pale eggs of pied wagtails, and speckled, greenish-brown eggs of reed warblers. Then they waded into the soggy fens surrounding Cambridge, England, and placed the models into reed warblers' nests. As long as the biologists introduced the plastic eggs during the afternoons of the hosts' egg-laying period, the hosts accepted all the speckled, greenish-brown eggs that resembled their own. They ejected all the other foreign eggs from their nests, and either abandoned those nests or built new ones atop the original nests.

Notice the condition: as long as the eggs were introduced during the afternoons of the egg-laying period. Reed warblers often rejected plastic eggs that the biologists introduced at dawn, the hour the birds normally lay their eggs. Perhaps they were most aware of the contents of their nests at that time. Brooke and Davies found that if they introduced the plastic eggs before the host had laid her own first egg, she always rejected the ringer as if she thought to herself, "I haven't laid any eggs yet. This must be a trick!" The investigators also tried eggs that were larger than cuckoo-sized eggs; the hosts often rejected these. Brooke and Davies speculated that reed warblers lower their chance of being parasitized by discriminating eggs on the basis of color, timing, and size. The cuckoos increase their chances of successful parasitism by mimicking the egg color and by laying eggs in the afternoon. Furthermore, the cuckoos have evolved an egg size of just the right dimensions to avoid discrimination by their hosts.

Not all hosts discriminate against cuckoo eggs as well as do reed warblers. Dunnocks, brownish sparrow-like birds often called "hedge-sparrows," lay unspotted turquoise eggs but do not reject the speckled cuckoos' eggs. They will even incubate eggs that are all black or all white. Brooke and Davies suggested that the dunnock may be a fairly recent host for cuckoos and has not had enough evolutionary time to develop discriminatory behavior. Only about 2 percent of dunnocks in Britain are parasitized by cuckoos. Computer simulations suggest that about 3,000 generations (approximately 3,000 years) would be required for anti-cuckoo discriminatory behavior to evolve in the dunnock population.

Given that common European cuckoos have long been known to be aggressive and destructive of other birds, how do people view them? Moses warned the Israelites—in addition to do not kill, commit adultery, steal, and so on—not to eat the flesh of "unclean" birds such as the owl, night hawk, pelican, cormorant, cuckow (cuckoo), et cetera. (Bible, in Leviticus and Deuteronomy, King James Version). Nowadays some people view cuckoos as cheaters, just like cowbirds. And then there are those who have been cheated on—cuckolded. The word "cuckold," refers to a man whose wife has committed adultery. In *Othello,* Shakespeare has Desdemona saying, "Who would not make her husband a cuckold to make him a monarch?" On the other hand, many Europeans consider cuckoos harbingers of spring, and for the Danish, cuckoos are a symbol of fertility and longevity. Cuckoo clocks imitate the birds' simple, musical song, "*coo-coo, coo-coo.*" Beethoven showcased the cuckoo in the third movement of his Symphony No. 6 in F, "The Pastoral," in which the woodwinds play a cadenza featuring the songs of the nightingale, quail, and cuckoo. Perhaps fortunately for them, these brood parasites have some redeeming qualities.

BIRDS AREN'T THE only parental deadbeats. Some ants pawn parental care off on to other ant species. One social parasite is the rare *Teleutomyrmex schneideri* ant, found in the Swiss and French Alps, living exclusively with another ant, *Tetramorium caespitum*. Pairs of *T. schneideri* mate inside the host ants' nests. After mating, the queen either sheds her wings and lays her eggs in that nest, or she flies out and searches for a different *T. caespitum* nest in which to lay her eggs.

Because *T. schneideri* lacks a worker caste, queens must rely on their hosts to care for them. Why don't the *T. caespitum* hosts evict the invaders? As pointed out by Bert Hölldobler and Edward O. Wilson, "Ants are easily fooled." Other organisms can break the communication code and exploit ants' elaborate social system. In Hölldobler and Wilson's words, "The social parasites that accomplish this feat are like burglars who enter a house quietly by punching the correct four or five numbers to turn off the alarm system." By using chemical signals, *T. schneideri* invaders fool the host ants into accepting them as full colony members. The host workers lick the invader queens frequently and regurgitate food for them. The tiny *T. schneideri* queens, one-tenth of an inch in length, do nothing for the host colony. But they certainly take from their hosts. They ride piggyback on the host colony queen and leave their offspring's care to the host workers.

Other cases of social parasitism in ants involve enslavement. Five species of Amazon ants—large, shiny red or jet black ants living in Europe, North America, the former Soviet Union, and Japan—use slaves to feed, clean, and move their young around, to feed themselves, and to excavate their nests. Amazons depend entirely on slaves that hatch from cocoons they steal from colonies of similar-looking ants in the genus *Formica*. While in their own slave-dug nest, Amazon ants stand around idly or beg food from the slaves. Outside the nest, they turn pugnacious. Fighting other ants and raiding for slaves are the only activities these ants do well.

Hölldobler and Wilson describe an Amazon ant raid as a spectacular event.

Workers pour out of the nest to form a compact column running over the ground at 3 centimeters a second—the equivalent of a human brigade traveling at 26 kilometers (16 miles) an hour. When they reach their target, a nest of Formica ants, they charge into the entrance without hesitation, seize the cocoon-covered pupae, speed out again, and return to their own nest. They attack and kill any worker that opposes them, piercing the heads and bodies of the defenders with their saber-shaped mandibles. Once home, they turn the pupae over to the adult slaves for further care, and revert to their usual indolence.

The slaves carry out their nursemaid duties acting as if they were natural sisters of the slave-makers. They faithfully perform the same tasks they would if they were home in their own colony. Evolutionary development has programmed them to do so, regardless of the context.

SOME FISHES SCAM other fishes into foster parenting their young. Cichlid fishes, or "mouthbrooders," protect their large, yolky eggs in their mouths. After hatching, the fry remain in their mother's mouth while they absorb their yolk. In some species, once the yolk is gone, the young fish swim out to forage for food but scurry homeward at the first inkling of danger.

A catfish from Lake Tanganyika in Zaire, Africa, exploits the highly advanced parental care behavior of at least six species of cichlid fishes. The catfish presumably sneak in and deposit their own eggs in the cichlids' mouths as the cichlids spawn. The mother cichlids incubate both their eggs and the foreigners' eggs. Unfortunately for the cichlids' own offspring, the catfish eggs hatch first. The catfish fry absorb their yolk within three days and turn on newly hatched cichlid fry. Smaller catfish fry bite the host fry and suck up the yolk. Larger catfish fry behave even more aggressively. They swallow baby cichlids in one gulp. What a thank-you for tender loving care by a foster mother! Typically a host female cichlid ends up brooding 1 to 8 catfish young and many fewer than her original complement of up to 50 of her own, as many of her young have served as snacks for the catfish fry.

Another example of brood parasitism involves a freshwater perch from Japan and the southern end of the Korean Peninsula. Male perches establish three-foot diameter breeding territories from which they aggressively repel intruders. Within these territories, the males clean the surface of several reed stems—future egg-laying sites for visiting females. Once they become fathers, the males defend their eggs and fry against predators.

A species of minnow takes advantage of this perch's parental care system by entering the perches' territories and depositing its own adhesive eggs on the cleaned reed stems. The perches' response? If a single pair of minnows invades a territory, the resident male perch drives them away. But the minnows often spawn in large groups, overwhelm the male perch, and lay their eggs on his cleaned reed stems with impunity. The father perch aggressively defends his nest site whether or not it includes minnow eggs in addition to his own eggs. This protection is critical for both species. When father perches are removed from their territories, other fish consume all host and brood parasite's eggs within two hours.

Obviously the minnows cash in on a good deal, but does their parasitic behavior hurt the host perch? Yes, but in an unexpected way. A third species of fish, the dark chub, swims along with and resembles the minnows— both the chubs and the minnows have a thick black stripe running down their sides. Dark chubs try to invade the perches' territories and eat their eggs. If the dark chub comes by itself, the perch ousts it. When spawning minnows are present, though, the male perch does not repel chubs, perhaps because he is confused and overwhelmed by so many black-striped invaders. The upshot is that perches' egg clutches in territories parasitized by minnows have much lower hatching success than those in territories not parasitized by minnows.

FROM THE HUMAN STANDPOINT, brood parasitism may seem like cheating, but "the wisdom of Nature" is more profound and complex than our opinions. We can only speculate how such behavior evolved. For a given species, at first brood parasitism was probably sporadic or even accidental, with only a few individuals in a population pawning off their offspring on their neighbors. Through time, however, if those parasitic individuals reaped greater reproductive success than others that laboriously cared for their own young, natural selection would have selected for the "shiftless" behavior. The rest is history, resulting in species totally incapable of caring for their offspring.

DEFENSE CONTRACTS

We could see the head and back of an alligator in the pond, and at one edge her nest stood, a brown mound against the gray-green marsh. The patch of willows curved in a thin fringe around the nest and pool and the herons were nesting in the trees. . . . Partly they were there because the only trees in the marsh were those willows that had found foothold in the spoil pile thrown up by the work of the alligator and her ancestors. But another factor in their presence was surely that the egg eaters of the marsh found it nerve-racking to rob [birds'] nests over an alligator hole.

ARCHIE CARR, *A Naturalist in Florida*

"Yikes!" I yelped as the prehistoric-looking reptile lumbered toward me. It was my first year teaching at the University of Florida, and I was exploring nearby Paynes Prairie. As I watched several male red-winged blackbirds defending their territories, nearby rustling noises startled me. The grass parted and revealed a huge alligator about 45 feet away. I had inadvertently gotten too close to her nest, and she was mad.

I mumbled an apology and backed up. She obviously didn't want to venture too far from her nest and apparently decided I wasn't a threat after all, though I doubt my apology had much of an effect. We left each other in peace, but I immediately understood why birds frequently nest near active alligator nests.

Female alligators heap vegetation, plant debris, and mud into mounds three feet high and six or seven feet across, then lay their eggs inside. Heat from the sun and rotting vegetation fuels the incubators. These nests provide optimal conditions for other reptiles' eggs as well. Mud turtles, red-bellied slider turtles, Florida softshell turtles, and Carolina anole lizards sometimes lay their eggs inside gator nests. An added benefit for these reptiles is protection from raccoons and other predators while mother gators aggressively defend their own eggs.

We'll look at four ways that animals gain protection from other animals' defenses.

FIRST, AS WITH Archie Carr's herons nesting near the alligator nest, some animals live near (or inside) other animals with great defenses. Many beetles and other insects live in ants' nests. Because they take on the colony scent, the ants don't recognize them as foreign. The insects gain protection from predators simply by associating with worker ants that vigorously defend the nest. William Wheeler noted how striking these relationships are because to us, at least, the foreigners are so different from ants. In his 1923 book *Social Life among the Insects*, Wheeler writes: "Were we to behave in an analogous manner we should live in a truly Alice-in-Wonderland society. We should delight in keeping porcupines, alligators, lobsters, etc., in our homes, insist on their sitting down to table with us and feed them so solicitously with spoon-victuals that our children would either perish of neglect or grow up as hopeless rhachitics" ("rhachitics" [*sic*] = rachitic = affected with rickets).

The tentacles of sea anemones are studded with stinging cells that contain minute harpoon-like capsules called "nematocysts." When a predator or potential prey touches a tentacle, the nematocysts discharge and tiny daggers inject a potent nerve toxin. Clownfish—like the star of the animated movie *Finding Nemo*—hide out among anemone tentacles. Bright orange with white bands or stripes, the bodies of these fish are covered with a special mucus that "fools" anemones into failing to discharge their nematocysts when touched. Simply by associating with sea anemone's armed tentacles, clownfish gain protection from their enemies. Both fish and anemone benefit when each eats food left behind by the other.

Some birds build their nests near the nests of more aggressive birds. For example, azure-winged magpies often build their nests close to those of Japanese lesser sparrowhawks. Mutsuyuki Ueta found that nearly 92 percent of lesser sparrowhawk nests in his Tokyo study site had magpie nests within 150 feet. The hawks drive off jungle crows, common egg and nestling predators. When hawk nests were nearby, magpies didn't waste energy defending their nests, and they fledged more offspring than did magpies nesting farther away from hawks. The hawks ate a magpie from time to time but not often: magpies made up only 0.5 percent of the hawks' prey. Not much risk, considering the substantial benefit.

Some birds that exploit other birds' defenses are social parasites. Along the Manú River in southeastern Peru, several waterbird species often nest together. Martha Groom found that a given beach might have from 2 to 200 nests of sand-colored nighthawks mixed with 1 to 12 nests total of black skimmers, large-billed terns, and yellow-billed terns. All four species laid their eggs in shallow depressions on dry sand. The nighthawks never defended their nests against predators. In contrast, the aggressive skimmers and terns defended the airspace or beach for as far as 300 feet from their nests, mobbing predators such as hawks, black caracaras, and bat falcons. Thus, skimmers and terns indirectly defended nearby nighthawk nests. Nighthawks with nests within 30 feet of these aggressive birds hatched more young than those with nests more than 90 feet away.

Clearly nighthawks benefited from the presence of terns and skimmers, but these other birds suffered. Large concentrations of nesting nighthawks attracted more predators, which meant that the defenders spent more time chasing and mobbing predators and less time caring for their young. Both species of terns fledged significantly fewer young from nests surrounded by many nighthawks. Why do terns and skimmers put up with nighthawks? Skimmers and terns are the first to arrive at the beaches. Several days later, the nighthawks arrive and begin laying their eggs. The only way for the early arrivals to avoid them would be to abandon their own eggs and go elsewhere to nest.

SOME ANIMALS INGEST other animals' defenses. Poison dart frogs use chemical defense to repel predators. These frogs earned their name because several groups of Indians from South America use skin secretions of certain species to poison their blowgun darts. The secretions contain potent alkaloids (a class of chemical compounds) that cause heart failure in animals that eat the frogs. Poison dart frogs that harbor these toxins are brightly colored blue, purple, orange, yellow, or red as a warning to potential predators: "Don't

eat me! I'm poisonous!" Individual predators either learn to avoid these frogs, or through time predator species evolve recognition of warning colors.

The most toxic frog known is the 1½- to 2-inch long golden poison dart frog, *Phyllobates terribilis,* from Colombia. One of these bright yellow, yellow-orange, or pale metallic green frogs has enough poison to kill 10 adult humans or 20,000 mice. A person could die from handling one of these frogs if a tiny amount of poison entered the bloodstream through an open wound. Indians in the southern Chocó of Colombia use these frogs' secretions to poison their darts—carefully, by protecting their hands with leaves as they rub darts across the frogs' backs. Why don't people die from eating game killed from these darts? The toxins are destroyed by heat. Monkey sushi wouldn't be a good idea, though.

Biologists once assumed that poison dart frogs produced their alkaloids through metabolic pathways. Observations of poison dart frogs raised in captivity, however, revealed that captives don't develop skin toxins, suggesting that the frogs get alkaloids from their food. In the early 1990s, the late John Daly and his colleagues experimented with green and black poison dart frogs raised from tadpoles collected in Panama. Frogs fed only wingless fruit flies did not develop skin toxins, whereas those fed arthropods collected from Panamanian leaf litter developed high concentrations of several alkaloids. The authors analyzed arthropods fed to the frogs and determined that certain alkaloids found in the frogs' skin almost certainly originated from beetles and small millipedes. This and other experiments strongly suggest that poison dart frogs incorporate their prey's alkaloids into their own skins.

Ants also produce large amounts of alkaloids. The most toxic poison dart frogs eat huge quantities of ants. In contrast, species of *Colostethus* from the same family that eat few ants aren't brightly colored and don't produce skin toxins. Ants were considered to be the main source of alkaloids for poison dart frogs until a recent paper revealed the importance of orabatid mites.

Also called "beetle mites," orabatids have hard, shiny shells that resemble black or dark brown beetles. Most are tiny—less than ½₅ of an inch long. These mites are among the most abundant arthropods living in soil and leaf litter in both temperate and tropical areas. A 1991 study showed that although mites make up a relatively small fraction of the diet of most frogs, some poison dart frogs specialize in eating them. In fact, the frogs eat mites in higher proportion than the mites' occurrence in the frogs' foraging areas. Mites are abundant and slow-moving, which makes them easy prey. But they contain lots of indigestible chitin. Why, then, do these poison dart frogs specialize in eating mites? Ralph Saporito and his colleagues recently provided a likely explanation. Alkaloids.

Saporito and colleagues extracted about 80 alkaloids from orabatid mites collected throughout Costa Rica and Panama. Presumably these alkaloids provide chemical defense for the mites against their predators, though obviously they don't protect them from poison dart frogs. Forty-one of the mites' alkaloids also occur in strawberry poison dart frogs, flashy red frogs with bright blue legs. These frogs feed heavily on orabatid mites. The authors concluded that the mites are a major dietary source of a wide variety of alkaloids for the frogs, and no doubt for many other mite-eating poison dart frogs as well. No wonder strawberry poison dart frogs specialize in eating ants and mites. By sequestering their prey's alkaloids, they gain protection against their own predators.

Nudibranchs, naked marine snails known as sea slugs, also sport bright colors warning predators that they taste bad. Some sea slugs have another defense, which they get from hydroids—small colonial animals with stinging cells armed with nematocysts like those of sea anemones. As in sea anemones, when a predator touches a hydroid, the nematocysts discharge and tiny daggers inject toxin. When nudibranchs eat hydroids, however, the nematocysts don't discharge. Instead, the stinging cells pass through the sea slug's digestive system and eventually collect in special-ized sacs that open to the outside. When a predator touches the sea slug, the stored nematocysts discharge and sting the attacker.

SOME ANIMALS USE other animals for their own defense by wield-ing them as shields. The following are three examples from the marine world. Hermit crabs wedge themselves into empty snail shells to protect their soft abdomens. Some, as an added protection, camouflage their shell homes by nudging sea anemones loose from rocks and lifting them onto their shells. Instead of looking like tasty crabs, now they appear as venomous anemones. If an octopus, squid, or other predator touches the crab, it gets stung by the anemone's discharging nematocysts. Both parties benefit. The crab gains protection, and the anemone gets a free ride to food. These hermit crabs are excellent

scavengers and quick to find dinner. As the crab tears into its meal, the attached anemone feasts on tidbits floating about in the water. The association probably allows the anemone to eat more than if it were still attached to its rock, engulfing food swooshing by in the current.

As Daphne Fautin phrases it: "Crawl like a crab, sting like an anemone." Boxer crabs pry anemones loose from their substrates and grasp one sea anemone in each slender claw. They cruise through their marine surroundings, resembling street boxers primed for a fight. If an attacker approaches, the boxer crab lunges and threatens with its "gloves," and if the potential predator persists, it gets punched with discharging nematocysts. A female boxer crab brooding tiny eggs under her abdomen protects not only herself, but also her young thanks to the stinging anemones.

Sponge crabs cover themselves with pieces of live sponges, which stick to stiff hairs on the crabs' carapaces. The skeleton of most sponges consists of tiny, needle-like spicules composed either of calcium carbonate (limestone) or silica, the mineral used to make glass. Fishes and other crab predators generally don't eat sponges because of this skeleton. Thus, the spicules that protect sponges also protect sponge crabs.

FINALLY, SOME ANIMALS offer food in exchange for the protection they receive from other animals. During the spring, summer, or fall, look closely at the flowers, weeds, or shrubs you encounter in your garden or in the wild. Some will surely sport aphids (plant lice) attended by ants. Watch for a few minutes and you'll likely see an ant touch an aphid with her forelegs or antennae. The aphid responds by extruding a drop of excrement from its anus—sugary liquid called "honeydew," a nutritious cocktail the aphid excretes from the phloem sap it slurps up from plants. The ants protect their sugar faucets by driving off parasitic wasps and flies intent on injecting their eggs into the aphids' succulent bodies, and by repelling beetles and other predators prowling the vegetation. Scale insects, mealybugs, froghoppers, leafhoppers, and treehoppers also provide ants with sugary secretions. Again, the ants guard their sugar sources against predators.

A variation on this association has evolved in caterpillars of certain lycaenid butterflies. Because these caterpillars eat plant tissue rather than phloem sap, they produce cellulose-laden feces of no interest to ants. Instead, the caterpillars have two types of glands that attract ant bodyguards. Pore cupola glands, scattered across the caterpillar's body, contain chemicals that attract ants like magnets. The Newcomer's gland, located near the caterpillar's rear end, secretes an enriched sugar solution—food for ants. As in their relationship with aphids, the ants guard these sugar sources by protecting the caterpillars from parasitic flies and wasps that lay eggs in and on caterpillars, and from predacious ants and wasps. When searching for plants on which to lay their eggs, some female lycaenids actively seek plants that already have ants, ensuring that their young will have solicitous attendants when they hatch. Ant bodyguards are critical for some species: unattended caterpillars likely get eaten or parasitized.

MOST LIKELY PUNKS who dye their hair brilliant pink, green, blue, or purple don't eat ants or orabatid mites, so they're no more poisonous than anyone else—certainly much less so than some CEOs of lending institutions. And being the most noisy and obnoxious species around, we can't do as sand-colored nighthawks do, nor has using porcupines as living shields ever really caught on. Still, the next time you walk by someone's yard and a rottweiler or German shepherd scares the living daylights out of you, think about this: Is the dog first and foremost protecting its owner, or is the owner just a mud turtle taking advantage of the dog's fierce defense of its own territory?

COW PIE NO. 5

Dogs will often seek out the ripest, most putrid, most god-awful things . . . and then with every sign of acute pleasure pull their lips back much like a horse smelling a mare in heat, and . . . do a shoulder roll right into the middle of the mess: they bend their forelegs and repeatedly rub the side of the neck and the top of the head into the object, sometimes switching sides and then finally rolling over onto their back and wriggling over the spot.

STEPHEN BUDIANSKY, *The Truth about Dogs*

Today my long-haired dachshund rolled in something really foul in the woods. Conan's usuals are elk urine and peccary poop, but this was something stronger . . . maybe a dead squirrel?

Why do dogs instinctively roll in stench? By masking its own scent, a dog might sneak up on prey without being detected. Dogs' ancestors are, after all, wolves—and wolves routinely roll in carcasses and other animals' excrement. Another theory is that wolves roll in carcasses to deliver the scent to the pack, as a way of saying, "Look what I found! Dinner!" Dogs might have inherited this tendency from wolves, although the behavior now elicits scolding and dreaded baths from their human companions. A third idea: Some experts suggest that by masking their own scent and taking on "unique" scents, dogs can attract more attention from other dogs.

Dogs and wolves aren't the only canids that self-anoint with stench. Foxes and coyotes also roll in carcasses, twisting back and forth to cover their fur with scent. Some non-canids smear chemicals from other animals on their bodies as well.

YOUNG KOMODO DRAGONS rub their bellies in the intestinal contents of deer, boar, or other prey victims. They also pull wads of hair from rotting carcasses with their front claws and then rub their bellies and sides in the hair. Why? Perhaps to avoid being eaten. Their elders, the world's largest lizards at 10 feet and 200 pounds, cannibalize young Komodo dragons. By masking their own scent with that of partially digested vegetation or hair, the youngsters might live to see another day, since adults usually don't eat their prey's hair or intestinal contents. Adult Komodo dragons don't seem to engage in this odiferous behavior, lending support to this speculation.

NERVOUSLY THUMPING ITS feet and shaking its tail, a Siberian chipmunk approaches a dead snake. Once the chipmunk senses no movement, it gnaws on the carcass, then wipes bits of snake skin onto its body with its front paws. It might also rub snake urine and feces over its fur.

Tomomichi Kobayashi and Munetaka Watanabe found that Siberian chipmunks smeared themselves from carcasses of all four species of snakes presented to them. In contrast, the chipmunks did not respond to carcasses, urine, or feces of lizards, frogs, birds, or mammals—even foxes and badgers, both of which eat rodents. The investigators suspected that by smelling like snakes, chipmunks might appear as non-food items to snakes, which largely detect prey by smell. In another experiment, the biologists offered Japanese rat snakes a choice of a dead house mouse basted with fresh rat snake urine and an untreated dead mouse. The rat snakes ate fewer of the urine-basted mice. Kobayashi and Watanabe speculated that chipmunk-eating mammals might also be deterred by snake scent on the rodents' fur.

Siberian chipmunks aren't the only rodents to take advantage of snake odor. Barbara Clucas and her colleagues recently reported that both California ground squirrels and rock squirrels chew shed rattlesnake skins, then twist to the side and apply the scent to their flanks by licking their fur. They grab their tails with their forepaws and then lick along the tail from the base to the tip. Secondarily, they smear a little scent onto their rear legs and occasionally to their genital areas, front legs, and heads. The authors concluded that this behavior most likely serves to protect these ground squirrels from their main predators—the rattlesnakes themselves. By applying rattlesnake scent to their tails and rear ends, the ground squirrels might mask the scent of their own anal glands—the scent that makes them smell like ground squirrels. Another advantage of smearing the scent around the tail is that as the tail swishes, it would disseminate rattlesnake odor.

The investigators found that in both species, adult females and juveniles spent more time applying scent to themselves than did adult males. This difference makes sense. Juveniles are more likely to die from rattlesnake bite because their small size limits the amount of venom they can survive, and because they are less likely to escape an attack. Adult females are more susceptible than adult males to rattlesnake bite because they actively protect their young from the snakes and because they spend more time than males in giving alarm calls and thus exposing themselves to danger.

WHEN I LIVED in Brooklyn, New York, in the mid-1970s, a biologist friend, Butch Brodie, gave me a baby hedgehog for a pet. "Monsta" had been one of Butch's laboratory animals. One evening as I set Monsta on the carpet for his daily exploration, he licked my hand and then foamed at the mouth. He raised his prickly body on his stubby front legs, reached over his shoulder, and smeared white foamy lather onto his spines with his tongue. Contorting his body to the other side, he covered those spines with spittle. Although I knew about self-anointing in hedgehogs—from Butch's research with captives—I stared in disbelief. If I hadn't known about the bizarre behavior, I would have assumed my pet was dying. But Monsta was simply reacting to the residue of Spic and Span on my hand.

Hedgehog watchers have long known that these little mammals lick novel, smelly, or noxious substances, promptly froth at the mouth, and wipe the spittle onto their spines. But why? One investigator observed that adult wild hedgehogs self-anoint only during the breeding season, suggesting the behavior might be a sexual attraction signal. He also found

that a suckling hedgehog removed from its nest and placed in strange surroundings usually anoints itself, perhaps fabricating a distress signal
to attract its mother's attention. But there's more to the answer.

Butch wondered: Since hedgehogs and toads live in the same environ-
ments, might hedgehogs anoint themselves with toad poison, and if so,
would the poison make their spines more painful to whatever they jabbed?
When Butch confined hedgehogs and toads together, the hedgehogs
chewed the large poison-filled glands on the toads' heads, foamed at the
mouth, then smeared saliva mixed with toad poison over their spines.

To determine whether spines anointed with toad poison were more
potent than untreated spines, Butch and six graduate students jabbed
themselves in the inner forearm with spines from four treatments:
(1) spines washed in alcohol; (2) spines washed in alcohol, then coated
with hedgehog saliva; (3) spines washed in alcohol, then coated with toad
poison; and (4) spines removed from a hedgehog after it had anointed
itself with toad poison. They blind-tested the four spines in random order.
Only one person reacted to spines of groups 1 and 2, and then only with
reddened skin. In contrast, when poked with spines from group 3, six people experienced immediate intense burning and splotchy red areas around
the puncture sites. All seven people experienced intense burning and red
splotches when poked with spines removed from the hedgehog that had
anointed itself with toad poison. The burning sensation from treatments
3 and 4 lasted up to one hour.

Butch's experiments suggested that hedgehogs' self-anointing behavior improves their defense. When hedgehogs chew toxic toads, they steal
another animal's defense and use it for their own.

IF YOU'VE EVER handled a millipede, you know that these multi-
legged wonders protect themselves by releasing foul-smelling secretions,

including hydrogen cyanide and benzoqui-
nones. Several species of capuchin monkeys from the New
World tropics take advantage
of millipedes' secretions: the
monkeys rub these arthropods
over their fur. Ximena Valderrama and colleagues describe a
typical sequence for wedge-capped
capuchins from Venezuela: "Upon finding
a millipede, capuchin monkeys typically rub

it vigorously against the back and roll over it, while intermittently taking it in the mouth and slowly withdrawing it again. During mouthing they drool copiously and their eyes appear to glaze over." Capuchins often share millipedes. An individual approaches the user and attempts to take the millipede. If the user doesn't surrender it, the other monkey rubs its body and tail against the user. As many as four capuchins writhe against each other, applying millipede secretions onto their bodies. Some millipedes survive the molestation unscathed, but most end up bitten, torn apart, or decapitated. By biting or crushing the millipedes, the monkeys may release more of the defensive chemicals.

Wedge-capped capuchins slather themselves with millipede secretion only during the rainy season, perhaps not coincidentally the time when mosquitoes are out in full force. More than just an irritation, mosquitoes transfer parasitic botflies to the monkeys. When botfly maggots emerge from their subcutaneous cysts, they leave behind open sores and the risk of secondary infection. Valderrama and colleagues suggested that self-anointing with millipedes spreads benzoquinones, known to be strong insect repellents, over the monkeys' fur.

Three years after Valderrama and her colleagues published their paper, Paul Weldon and four colleagues reported on their experiments to test whether millipede benzoquinone does in fact deter mosquitoes. They placed silicone membranes laced with two different benzoquinone solutions in acetone and control membranes containing only acetone over wells filled with human blood, a preferred host for the yellow fever mosquitoes they used. Their results revealed that female mosquitoes landed less often and fed less frequently from blood under the benzoquinone-laced membranes than under the control membranes. Wedge-capped capuchins' "millipede sticks" seem to be effective mosquito repellents.

Weldon and colleagues also tested whether white-faced capuchin monkeys and tufted capuchin monkeys, both known to rub themselves with arthropods, would self-anoint when offered the two different benzoquinones. They treated filter papers with the two benzoquinones in acetone and with acetone alone, offered each monkey one of the three papers, and then recorded its behavior. None of the 22 monkeys self-anointed with the acetone-only papers. After a few seconds, they dropped what must have seemed useless pieces of junk. Thirteen of the 22 monkeys energetically wiped either one or both of the benzoquinone-treated filter papers against their fur and occasionally drooled.

Capuchins aren't the only primates that drool and whose eyes glaze over when self-anointing with millipedes. When black lemurs from Madagascar bite into millipedes prior to rubbing the wounded bodies onto their fur, they also salivate profusely and their eyes glaze over. To the human observer, they seem to experience blissful pleasure, as though millipede secretions were mind-altering drugs. Perhaps they are, at least to these non-human primates. I don't recommend that readers try it, however . . .

ANOTHER FREQUENT BEHAVIOR, this time among birds, would seem at first glance to be another example of anointing behavior with clear benefits. Over 200 species of birds—including starlings, tanagers, orioles, and weavers—grasp ants, caterpillars, millipedes, and wasps in their bills and rub the arthropods over their feathers. Starlings and blue jays pick up ants, place them under their wings, and allow them to crawl over their skin and feathers. Crows, jays, and magpies spread their wings over ant nests and allow ants to crawl over their bodies. All three of these behaviors are called "anting." Ants, wasps, millipedes, and many other arthropods produce substances that kill insects, mites, ticks, bacteria, and fungi. When these arthropods get riled, they release their chemicals by biting, stinging, or simply exuding the substances. By anting, do the birds get the arthropods to release these chemicals, which then might protect birds from parasites and microbes?

Most anting songbirds use worker ants of the subfamily Formicinae, which secrete formic acid. At certain concentrations, this corrosive acid kills bacteria and fungi. Recently, Hannah Revis and Deborah Wallen tested the effects of ant secretions and pure formic acid on two parasitic bacteria and three parasitic fungi that break down the structural integrity of birds' feathers. Surprisingly, none of the five species of ants they examined produced chemicals that inhibited the microbes tested. Although the ants secreted formic acid, the concentration was too low to inhibit the microbes. Pure formic acid, however, strongly inhibited all the bacteria and fungi tested.

Okay, so perhaps the explanation for anting isn't protection from microbes. Other suggestions include: Anting (1) removes stale lipids from the skin and feathers, (2) provides autoerotic stimulation, (3) stores ants in feathers for reserve food, (4) facilitates molting and soothes feathers, (5) removes formic acid and other chemicals from ants, making them more delectable for dinner, and (6) reduces the load of feather mites, ticks, and lice. The last hypothesis is the most widely accepted, but some studies have shown that anting doesn't decrease presence or abundance of feather parasites.

So, what's going on? Is anting a good thing to do, and we just haven't discovered why? Do birds simply enjoy being tickled? Anting is a mystery yet to be solved.

MEANWHILE, CONAN HAS just had a bath. That means he'll be searching for a new cologne to mask what he no doubt perceives as the foul stench of Johnson's baby shampoo. What will he find in the woods tomorrow? Cow Pie No. 5? Old Elk Spice? With Peccary Love? Revlon Coyote Flair? Supposedly, interest in rolling in smelly discoveries diminishes with a dog's age. Now I'm wondering if my deodorant consumption is going down as I approach Social Security age . . .

AUDACIOUS PIRATES AND SNEAKY BURGLARS

[The bald eagle] is a bird of bad moral character; he does not get his living honestly. You may have seen him perched on some dead tree, near the river where, too lazy to fish for himself, he watches the labour of the fishing hawk; and when that diligent bird has at length taken a fish, and is bearing it to his nest for the support of his mate and young ones, the bald eagle pursues him and takes it from him. With all this injustice he is never in good case; but, like those among men who live by sharping and robbing, he is generally poor, and often very lousy.

BENJAMIN FRANKLIN, AS QUOTED IN H. W. BRANDS,
The First American: The Life and Times of Benjamin Franklin

I first witnessed a bald eagle steal from a fishing hawk (osprey) in Gainesville, Florida, during a field trip with my vertebrate zoology students. We watched an osprey hover over Lake Alice, then plunge into the water in a spectacular head- and feet-first dive. When the bird surfaced and then flew upward with a fish clutched in its claws, we all cheered. I started to describe the piracy habits of bald eagles toward ospreys when a student yelled, "Look!" A mature bald eagle swooped down, and the osprey dropped its catch. The eagle snatched it in midair and flew off. Half of the students cheered for the pirate; the other half booed. It was a teaching moment I'll always remember.

Benjamin Franklin objected to using a thief and pirate as our national emblem for moral reasons. I wonder, though, if Franklin wasn't a bit too hard on bald eagles. Thievery is just another way for many animals to get a meal.

BIOLOGISTS CALL THE behavior of stealing food from other animals "kleptoparasitism," from the Greek *kleptes,* meaning "thief." Recall from the essay "Bubble Blowers" that a likely explanation for pack hunting in wolves is the ability to offset the loss by ravens stealing their prey. Animals steal both from their own species and from others, but here we'll look at only the latter.

Seabird nesting colonies provide rich pickings for pirates because an unending stream of parent seabirds carries food back to the colony for the chicks. Some birds carry fish in their bills, which makes the food clearly visible. Others carry food in their crop, a

sac-like structure at the base of the esophagus, and regurgitate the mass for their young once at the nest. Some thieves, such as frigatebirds, can tell whose crop is empty and whose is full. They chase only birds with distended crops.

Frigatebirds, with eight-foot wing spans and lightweight bodies, are magnificent flying machines. Often called "Man-o'-War Birds" or "Pirates of the Sea," these giants soar over nesting colonies of other seabirds scoping out theft possibilities. The red-footed booby—a comical bird that shuffles about on short legs and fully webbed feet the color of ripe tomatoes—is a frequent victim. In the 1960s, Bryon Nelson spent a year on two uninhabited islands of the Galápagos and frequently watched frigatebirds steal food from red-footed boobies in flight: "A booby hemmed in by several frigates and stubbornly refusing to throw up was seized by the tail or wing tip and capsized. This usually encouraged him to give in and he then began to regurgitate, pointing his bill downwards to aid the process. The frigates snatched eagerly at the fish as soon as the slimy bolus appeared, catching it in the air or following it down to the sea."

In the bird world, turnabout seems fair play. You steal, you get stolen from. Frigatebirds have a tough time feeding the stolen goods to their own chicks because laughing gulls try to snatch food from the chicks' gullets.

Although this technique rarely works, the gulls do sometimes surprise frigatebird chicks and cause them to drop their dinners.

Laughing gulls victimize many other seabirds, including terns. Once a laughing gull starts to pursue a tern in flight carrying a fish, other gulls may join in. As many as eight laughing gulls chase the tern in single file. Zigzagging through the air, the tern can often outmaneuver the gulls in the lead. But then, from nowhere, a gull from the back of the line cuts a corner and heads off the tern. If the tern doesn't drop its catch, the gull might grab the fish directly from the tern's bill. Groups of gull pirates often are more successful than loners, but since a tern carries only one fish at a time, only one gull benefits from the chase.

Alan Burger studied kleptoparasitism on Marion Island in the sub-Antarctic. Rockhopper penguin chicks call excitedly when their parents return to the nest with food. Their calling attracts lesser sheathbills, small white shorebirds that superficially resemble domestic pigeons. As the penguin parent passes a mass of regurgitated crustaceans, squid, or fish to its chick, a lesser sheathbill crashes into the adult. The collision may cause the penguin to spill food. If so, the sheathbill snatches the gob and flies off to give it to its own chicks. Sheathbills also peck at the penguin parents until they drop their food. Although most avian pirates gain only a small fraction of their total food through piracy, the sheathbills on Marion Island got most of theirs by stealing it. Sheathbill parents may not have been able to raise their chicks without the stolen goods.

In some interactions, such as that between sheathbills and penguins, only about 2 percent of potential victims get robbed. Because penguins far outnumber sheathbills, a victim's chance of getting robbed more than once is small. Some thieves, such as gulls, may do more damage. One study found that silver gulls stole 28 percent of the fish that lesser-crested terns carried back for their chicks. Another study reported that lesser black-back gulls and Atlantic puffins stole one-third of the food that herring gulls carried.

Some avian victims chase thieves or even fight back. Other victims ignore the thieves. The response may depend in part on how easily a victim can catch another meal. If food is abundant, the reaction might be to shrug off the encounter. The response might also depend on the size difference or difference in aggression levels between the birds. Just like the shy kindergarten boy who chooses to walk away from a sixth-grade bully, bird victims may choose when to fight back and when it's not worth the risk of getting hurt.

MANY INSECTS ALSO steal food, but their morality hasn't been scrutinized because they never had the slightest chance of becoming our national emblem. Consider ants that live in the southwest U.S. deserts. Harvester ants gather mostly seeds, which they store in underground nests. But they also collect dead insects, especially termites. Honeypot ants often stop harvester ants returning to their nest. The honeypots climb onto the harvester ants' backs. They nibble the workers' heads, mandibles, and mouthparts for a few seconds. Harvester ants carrying seeds usually tolerate the inspection and then proceed homeward. If the harvesters have dead insects, however, the honeypots grab the booty. Honeypots often gain much of their insect protein through this highway robbery.

All over the world, ants have "struck a deal" with sap-sucking insects such as aphids (see "Defense Contracts"). Aphids pierce plant leaves, flowers, stems, and roots to feed on sap. These tiny insects excrete the

excess carbohydrates as the sweet liquid honeydew. Ants adore honeydew, which they slurp from aphids' rear ends. Ants often tap aphids with their antennae, the signal they want to be fed. The aphids oblige by releasing drops of honeydew. Why the generosity? Ants protect their sugar faucets—aphids—by chasing away predators such as ladybugs and parasitic wasps and flies. It's an arrangement that benefits both parties.

But in some parts of the world, ants don't get to keep all of their hard-earned protection money. Mosquitoes of one Oriental and African genus haunt trees where ants run along the trunks carrying honeydew back to their nest. A mosquito lands in front of an ant and vibrates its wings. As soon as the ant opens its jaws, the mosquito pushes its proboscis into the ant's mouth and slurps up the honeydew. These mosquitoes have a modified proboscis, perfect for filching droplets. In fact, honeydew is probably the only food they ever eat. Why might the ant give up its honeydew without a fight? As highly evolved social insects, worker ants feed other ants (see "An Intimate Act"). Food-gathering workers have two stomachs, the larger one being the crop. Ingested food dissolves into a liquid, which is stored in the crop. A hungry nestmate strokes a food-gathering ant's antennae. The ants put their mouths together. The food supplier upchucks some liquid into the other's mouth. Perhaps the giving of honeydew by worker ants to mosquitoes is just an extension of ants' innate sharing behavior. The mosquitoes' wing-vibrating behavior appears to allow them to exploit the ants' social behavior.

A species of chloropid fly rests by day near orb-weaving spiders that hide in curled leaves off their webs. As soon as an unsuspecting insect crashes into the web, the spider charges out, injects venom, and wraps the insect in silk. The spider carries its treasure back to its retreat. The waiting fly inches toward the spider. Once close enough, it unfolds its long sucking proboscis. It inserts the proboscis into the wrapped prey and sucks. After a minute or two, the sated fly waddles away, leaving the spider with the dregs.

Tiny jackal flies steal from certain orb-weaving spiders that sit in the middle of their webs during the day. The flies often perch on the spider itself—a good way to avoid a brush-off. A large orb-weaving spider might have eight jackal flies on its abdomen. After the spider has injected venom into a victim and the prey's body has been partially digested, the flies scoot off the spider, slurp a snack, and return to their perch. Some jackal flies steal bits of food from crab spiders and predatory insects. Others lick pollen that worker bees have collected on their legs. Still others do the mosquito trick of stealing honeydew from ants.

SOME MAMMALS, BIRDS, insects, and other animals sneak into other animals' homes and burgle their food.

Many animals hoard edibles for times when food may be hard to find. Harvester ants store seeds in underground nests. Chipmunks carry seeds, nuts, and bulbs in their cheek pouches and deposit the items in larders in their nests. Moles store earthworms in their tunnels. Some owls construct larders at their nests where they store excess rodents and birds they've killed. Another form of hoarding is laying in provisions for the next generation. Some female wasps sting and paralyze spiders or insects, which they store in their nests. The wasp lays an egg on the prey and seals up the nest chamber. When the larva hatches, it has plenty of food while maturing (see "Be It Ever So Humble"). Bees store pollen moistened with nectar as food for their larvae.

Although food caches are critical for surviving lean times, the stockpiles attract burglars. Some animals break into a cache and carry off food. Rodents scurry away with seeds stolen from other rodents and harvester ants. Other burglars are live-ins: they eat the food while in the cache site itself. Earwigs, beetles, and other small arthropods infest and consume the larders of mammals and birds. Some arthropods lay their eggs on the stored provisions of bees, wasps, and beetles. After hatching, their young eat the provisions meant for their hosts' larvae.

The impact of burglars on food caches can be substantial. In one study, researchers artificially cached clams to simulate caches made by northwestern crows. Over 80 percent of the clams disappeared within seven days. In another study, researchers cached horse chestnuts. Within five months, gray squirrels stole nearly 85 percent of the nuts. In a study of satellite flies, so named because they closely follow wasps carrying prey back to their nests, investigators found that the flies laid their eggs in 31 percent of the digger wasp cells examined. In another study, parasitic anthophorid bees laid eggs in 59 percent of the provisioned nest cells of another species of anthophorid bee.

Some hoarders actively protect their stored food. Red-headed woodpeckers, giant kangaroo rats, and eastern chipmunks aggressively defend their caches. Worker honey bees sting to defend their hives from wasps and other honey thieves, including humans. Beetles attempting to rob honey from southeast Asian honey bees get mobbed and ousted from the comb.

Non-aggressive behaviors reduce theft as well. Many hoarding animals "squirrel away" their food in concealed sites, such as crevices of tree branches. Solitary wasps and bees plug their nest entrances with soil to deter ants. When hoarding food, animals are often secretive to avoid attracting attention. Some cover their caches with leaf litter, soil, or other debris, or hide their food stores underground.

KLEPTOPARASITISM, WIDESPREAD THROUGHOUT the animal kingdom, may seem unusual, but it's just part of nature. Who are we to label the behavior good or bad, right or wrong?

Franklin didn't admire kleptoparasites. But did he have a better choice for our national emblem? The turkey: "He is, though a little vain and silly, it is true, but not the worse emblem for that, a bird of courage, and would not hesitate to attack a grenadier of the British Guards who should presume to invade his farm yard with a *red* coat on."

Just think—if the vote had been different, we might feast on our national emblem every Thanksgiving!

3 Green, Green Plants of Home, & Other Interactions between Animals and Plants

ANIMALS DON'T INTERACT only with other animals, of course. They also interact with plants in a mind-numbing variety of ways. Some interactions are familiar. The pet poison dart frogs in your terrarium inhale oxygen and exhale carbon dioxide, while the philodendron you planted for the frogs to climb on "inhales" carbon dioxide and "exhales" oxygen when it photosynthesizes. A pair of robins builds their nest on the branch of a cherry tree outside your window, while very hungry tent caterpillars industriously munch the leaves off other branches. Later in the summer, the same robin may steal some cherries and drop their pits a fair distance away. Bees buzz around the flowers of the tomato plants in your garden, inadvertently pollinating them while harvesting pollen for their own offspring.

We want to focus on unusual plant-animal interactions here, though. Let's start with the interaction that the buzzing bees and tomato flowers illustrate: pollination. More than 75 percent of flowering plant species depend on animals for pollination. Many of these plants offer their pollinators rewards, most often nectar or pollen. Others offer no reward; instead, they trick, trap, or deceive animals into pollinating them. Certain sexy-looking orchids trick horny insects into "mating" with their flowers, effecting pollination in the process. Now for seed dispersal. Many kinds of plants depend on animals to disperse their seeds. Birds—such as the cherry-stealing robin—and bats are well known as fruit-eaters and inadvertent seed dispersers, but would you believe that the list also includes some fishes, frogs, lizards, and turtles?

Many animals depend on plants for shelter—and not simply as providers of nesting platforms. Some of the ways that animals use plants for homes are far from simple. Certain insects live in plants' ovaries and "stomachs." Weaver ants construct tent-like nests from silk in the treetops by using their own larvae as living shuttles. Tent-making bats modify leaves for shelters. Depending on the species, the tent might be only for a pair and its baby or for the harem male's entire entourage of females and their gaggle of offspring.

Humans are not the only animals to use plants for medicines, stimulants, and hallucinogens. If you were an African chimpanzee plagued by nodular worms, you might eat the scratchiest, hairiest leaves you could find. And if you were an Asian water buffalo forced to toil all day in fields of cultivated opium poppy . . . well, you just might graze on some of those flowers. Other animals rub citrus fruits onto themselves as insect repellents.

Biologists delight in discovering novel natural history relationships. Normally one expects that animals eat plants or that plants eat animals, as in the case of pitcher plants. But does one expect that some ants feed the plants that provide them with housing, or that some plants feed their ant houseguests that protect them against leaf-munchers?

SEXY ORCHIDS MAKE LOUSY LOVERS, AND OTHER ORCHID CONTRIVANCES

The more I study nature, the more I become impressed with ever-increasing force, that the contrivances and beautiful adaptations slowly acquired . . . transcend in an incomparable manner the contrivances and adaptations which the most fertile imagination of man could invent.

CHARLES DARWIN, *The Various Contrivances by which Orchids Are Fertilized by Insects*

We tend to think of orchids as flashy, colorful flowers carefully nurtured in elegant greenhouses by patient aficionados or purchased from florists to be worn as prom corsages. Many orchids, though, are small and not the least bit showy. Wild orchids grow on every continent except Antarctica. They range in size from tiny plants less than an inch high to 100-foot-long vines. Flowers come in all colors except black.

I'm not sure what it is, but there's something extra special about finding an orchid in nature. Large or small, flashy or demure, a wild orchid is always a treat. As a child, I admired pink lady's slippers in the damp woods near my home in the Adirondacks. Years later, in the cloud forest of Costa Rica, large purple orchids—the national flower—brightened my walk through an otherwise dark understory of dense vegetation. Along the road edges, orange and yellow orchids grew like weeds, mimicking tropical milkweeds. My favorites in Ecuador were carpets of large lavender orchids covering hillsides, and petite cream orchids clinging to tree branches. Elegant, one and all.

Look closely at an orchid and you'll appreciate these flowers' unique-
ness. The outermost part of most other flowers consists of green petal-like
sepals, which form the calyx. Orchids have three sepals that, instead of
being green, are often the same color as the flower's
three petals. Typically larger and more
showy than the other two, the central
petal, called the "labellum" or "lip,"
often boasts an unusual shape that includes
a landing platform for insect pollinators. In
flowers of most other kinds of plants, the
male organs (stamens) encircle the one
or more female organs (pistils). Most
orchids have only one stamen, which
is partially or completely joined with the
three pistils. Together they form the column,
which rises from the center of the flower. As any-

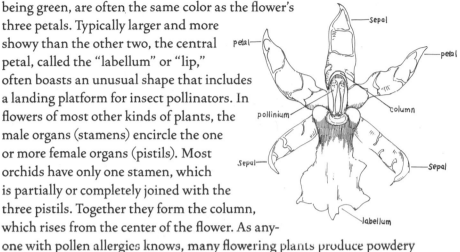

one with pollen allergies knows, many flowering plants produce powdery
pollen—that yellow stuff that coats your car and infiltrates your nose. Not
so with most orchids. Their pollen is clumped together in sacs (called "pollinia") attached to the stamen by appendages. The pollen sacs and appendages together are called the "pollinarium."

Darwin was impressed by the many ways orchids trick, trap, and
deceive insects to cross-pollinate them (transfer pollen from the stamen of
one plant to the pistil of another plant). According to Jana Jersáková and
her collaborators, "Darwin considered the adaptations of orchid flowers
to their animal pollinators as being among the best examples of his idea
of evolution through natural selection."

BUCKET ORCHIDS GROW on tree branches in wet tropical
forests of the New World. The central petal of each flower is
shaped like a round-bottomed jug or bucket. Once the flower
opens, two glands secrete liquid that drips into the bucket.
The flower exudes a fragrance that attracts metallic
green or golden yellow male euglossine, or orchid,
bees. As you'll see, the dripping liquid is neither
nectar nor the source of the irresistible scent, but
rather serves a different purpose.

Eric Hansen vividly describes the relationship
between the bucket orchid *Coryanthes speciosa* and
its male orchid bee pollinators as involving "an

ardent suitor, tantalizing promises, a noble quest, intoxicating perfumes, deception, dancing, and adventure." Bright yellow and yellow-green, the large flower of this orchid sports purple-red and rose spots and splashes. Metallic green male orchid bees swarm around the flower and try to land on a slippery structure that connects the bucket to the stem and front of the flower. If they get a foothold, the bees secrete lipids from glands associated with their mouths. Using mop-like brushes on their front legs, they mix the lipids with fragrant substances from the orchid flowers. The bees take off. While hovering, they brush the lipid-fragrance mixture into special pockets in their hind legs.

Biologists are not sure exactly how the male bees use their lipid-fragrance mixture. At one time it was thought that males use it as a precursor to producing their sex pheromones—love potions to woo females. In 2004 Benjamin Bembé suggested that males actively spray the stored fragrance from their leg pockets during courtship display. One component of courtship, called "ventilating," involves the male vibrating his hind wings. When he does this, comb-like structures on his hind wings hit the small brushes on his hind legs and spray a fine aerosol cloud of fragrance. Think of it as being analogous to rubbing your thumb across the surface of a toothbrush you've dipped in men's cologne. Mist the lady bee, and she can't resist.

Back to the *Coryanthes speciosa* orchid. Flowers of this species bloom for only a few days, and there's limited fragrance. Often many male orchid bees jostle into position, butting heads and shoving with their legs to gather perfume. During the scuffle, bees occasionally slip and fall or get knocked into the bucket. A narrow tunnel is the one and only escape route. Hansen describes the escape as follows:

Just at the entrance to this tunnel, on the inside wall of the bucket, is a step that the sodden male uses to climb out of the fluid and into the passageway. He then slowly squeezes and wiggles his way forward. But before he reaches open air, the bee must pass beneath a "twin pack" of pollinia—sacs containing thousands of pollen grains—situated at the roof of the tunnel, on the anther, the male part of the flower. At a precise moment, the pollen disengages and becomes lodged on the bee's back at the spot where the thorax and abdomen are hinged. By the time the bee has climbed free of the tunnel, the pollen is attached between his wings like a small backpack. Once out, the bee—wet and disoriented—pauses to dry himself on the flower's lateral sepals. His ordeal may have taken as long as forty-five minutes.

The flummoxed bee, still needing his perfume fix, flies to another *C. speciosa* blossom and again jostles his way in. If he falls into the bucket of the second flower, he'll pollinate it with the pollinia he's unwittingly carried from the first flower. As he negotiates his way through the exit tunnel, a catch mechanism on its roof grabs his pollen backpack, which lands on the pistil. "Wait a minute," you might be thinking. "What's the chance of one bee being such a klutz twice in a row?" Slim. Perhaps for this reason, pollination and subsequent formation of a seed pod happens infrequently in this orchid species.

THE *C. SPECIOSA* bucket orchid tricks and traps its bee pollinators, but at least it rewards them with perfume. Not all orchids are so generous. In fact, orchids are pros at deception. The family Orchidaceae, consisting of more than 20,000 species, is one of the largest in the plant kingdom. About one-third of all species offer their pollinators neither nectar nor any other reward. They dupe insects into pollinating them. Some exploit the normal foraging behavior of pollinators, others deceive horny male insects by mimicking sexy females, others dance in the breeze like intruders to be assaulted, and still others smell like something on which female insects lay their eggs.

EXPLOITING POLLINATORS' FORAGING behavior, the most common form of deception, has been reported in 38 genera. In these species, orchids that do not offer rewards closely resemble other, common non-orchid flowers that do offer nectar or pollen rewards. For example, they advertise through flower color, form, and fragrance. Insects arrive at the orchid flower "expecting" to find food, but instead they get trapped and zapped—with pollinia. If they are duped again later, they'll leave the pollinia behind, and the second orchid is pollinated. In the grass pink orchid of eastern United States and Canada, the erect lip is covered with bright yellow tufts of hairs that resemble a mass of closely packed stamens on similar-looking non-orchid flowers that often bloom nearby. Once a pollen-foraging bee lands on the lip, it is dropped onto the column by the hinged base of the lip—just like a trapdoor in a horror movie. In the process, it picks up pollinia and/or deposits pollinia from a previous misadventure onto the stigma.

SOME ORCHIDS ATTRACT male insects by mimicking the smell and shape of female insects. While attempting to mate with the deceptive flowers, the insects pick up or deposit pollinia. Unrelated orchids from Europe, Australia, Africa, and South America display this type of deception. A dragon orchid from Australia emits an odor similar to the pheromones of female thynnid wasps. Male wasps fly in from long distances, attracted by the scent. Once a male approaches, he sees the blossom's labellum, which resembles a female wasp. While attempting to copulate with the labellum, the male picks up pollinia on his head or abdomen. The next time he tries to copulate with a labellum, he deposits the pollinia from his body onto the sticky stigma.

Here's another good one, this time involving flies from the family Tachinidae in South America. When a female tachinid fly is ready to mate, she lands on a leaf in a sunny spot and signals passing males by opening and closing her genital opening. Sunlight reflects off her orifice, creating an irresistible sight for horny tachinid males. A male zooms in and copulates within seconds. The *Trichoceros antennifera* orchid mimics a female tachinid fly. The orchid's narrow column and lip base are barred yellow and red-brown and extend laterally, resembling a female fly sitting with extended wings. The flower's stigma reflects sunlight, looking like a female fly's provocative signal. A male strikes against the flower for as long as he would take to mate with a real female and inadvertently picks up the complete set of pollinia and associated parts. Then off to another "signaling female," where the pollinia is forced onto the stigma when he "mates" with that second flower. Clearly the tendency of males to think only with their gonads isn't limited to *Homo sapiens* . . .

ANOTHER TYPE OF DECEIT involves dancing orchids and territorial bees. Along the coast of Ecuador, male bees of the genus *Centris* staunchly defend territories by attacking and driving away invaders—other insects that fly by. Male bees rest on twigs or leaves, surveying their territories, ready to pounce on intruders. Certain *Oncidium* orchids have arched flower racemes (single flowers growing along a stem), whose blossoms dance in the slightest breeze. This movement induces aggressive male bees to attack and strike the flowers. Pollinia attach to the bees, and a continuing breeze keeps the irate bees striking blossom after blossom, pollinating the orchids in the process.

BLOWFLIES, FLESH FLIES, coffin flies, and other carrion flies lay their eggs in decomposing animals. When the larvae hatch, they have a ready source of gourmet food. Many carrion flies can detect the smell of dead bodies from over a mile away. Several thousand species of plants give off odors so similar to rotting flesh that they attract carrion flies. These plants dupe flies into pollinating them, but offer nothing in return. Flies that lay their eggs on these plants have thrown away their reproductive investment, as maggots that hatch from the flies' eggs starve.

Certain orchids use this type of deception. Greenhood orchids from Australia and New Zealand produce no nectar. And they stink, at least to the human nose. Superficially the pollination deceit in greenhood orchids and bucket orchids is similar: the insects cross-pollinate while escaping through tunnels. In other ways, these orchids are very different. Bucket orchids attract pollinators by smelling good to perfume-hunting bees; greenhood orchids attract by smelling good to carrion-seeking flies. The receptacles are formed from different structures, and the insects are "tricked" in different ways.

Connected at their base, the sepals and petals of greenhood orchids form a "kettle" that houses the column and the lip. Covering the kettle like a hood sit the interconnecting upper parts of the petals and sepals opposite the lip. When a fly lands on the lip, the structure tips over, throwing the insect against the stigma and trapping it in the lower chamber. As it escapes through a tunnel, the fly touches the pollinarium. When it visits another flower of the same species, it once again gets dumped into the chamber, and while escaping leaves the pollinia on the stigma.

Flies and gnats typically pollinate species of the orchid genus *Dracula*, meaning "little dragon" in reference to the face-like image with two "eye spots" in the flower's center. These Central and South American orchids have fungus-shaped lips that exude a fungus-like or fishy odor. Flies and gnats flock to the blossoms intending to lay eggs. As they move about, they cross-pollinate the flowers.

HOW IS IT possible for deceptive orchids to arise and persist? Here's one possibility. Certain insects strongly respond to particular stimuli, for example pollen-foraging bees to bright yellow stamens, male tachinid flies to genital openings, and gnats to fungi. If the deceptive orchids are sufficiently rare, and/or if they are relatively harmless (orchids often frustrate but don't kill the duped insects), perhaps the insects won't evolve recognition behaviors against the orchids. This could be because they generally find the correct stimulus—non-orchid flowers whose stamens are covered with powdery

pollen, female tachnidid flies opening and closing their genitalia, and fungi in which to lay eggs—or because making mistakes doesn't have that great a cost.

Deception promotes cross-pollination in orchids. The benefit to the orchids is efficient pollen movement between different individuals and higher seed quality. How so? Pollinators visit fewer flowers on an individual plant when they don't get rewards. They move along to other plants. From an orchid's perspective, all is fair in love and cross-pollination.

A SEEDY NEIGHBORHOOD

For a seed, getting dispersed by ants is like trying to get out of town on the local city bus. You often circle back to where you started.

ROBERT DUNN, "Jaws of Life"

Most plants need help dispersing their seeds. Some get help from gravity—they just let their seeds fall. Others cast easily airborne seeds to the wind. Aquatic plants often shed seeds that float on water currents. Many other plants take advantage of mobile animals to move their seeds.

Some animal-dispersed seeds, such as Spanish-needle and beggar's-tick, have hooks, spines, or barbs that catch on mammals' fur. Eventually the mammal grooms the seeds out or they fall off—often far from the parent plant. Plants benefit from their seeds being carried away from the parent plant because that way their offspring might experience less competition for light, water, and nutrients. Predation on the offspring might also be reduced if the seeds don't all land in one spot. Of course not all mammal-carried seeds end up on fertile soil. In the case of my long-haired dachshund, Conan, many of the numerous burrs and other seeds stuck to his fur end up on the carpet and between my sheets.

NOT ALL PLANTS get their seeds carried far from themselves. As pointed out in the quote by Dunn, seeds dispersed by ants often end up near the parent plant. Nevertheless, thousands of plants produce seeds that offer ants a white, fleshy reward, which encourages ants to disperse their seeds. This reward, a fatty edible appendage, is called an "elaiosome"— from the Greek *elaios*, meaning "oil," and *soma*, meaning "body." In addition to protein, sugar, starch, and vitamins, elaiosomes contain certain lipids that ants cannot get from other foods. Elaiosomes come in different forms: sheaths, caps, girdles, and finger-shaped extrusions on the tips of seeds. Worker ants grab seeds with their mandibles. Back in the nest, they eat the elaisomes and chew them into pieces to feed to their larvae. Workers toss the now elaiosome-free seeds into underground galleries and eventually carry many to the surface, where they dump the seeds into garbage pits and colony graveyards. With luck, the seeds germinate—surrounded by cricket legs, rival ants' heads, ant corpses, and other refuse that provide nutrients.

Even though ants might move seeds only short distances from where they fell to the ground, thousands of plants use this transport mode provided by a highly organized workforce. What's so good about ant dispersal? The answer might depend on the environment; thus, there might be different answers.

In North America, ants' nests are often more fertile than surrounding soils, richer in nitrogen and phosphorus. In Robert Dunn's words: "The grass is always greener over the septic tank." Nests contain discarded prey, droppings, and dead nestmates. Seedlings germinating from ant nests often are larger and have greater survivorship than seedlings from other areas. The "compost hypothesis" suggests that ant dispersal might provide seeds with fertile germination sites.

If this compost hypothesis has merit, plants growing in soils generally poor in nutrients should be especially likely to entrust their seeds to ants. In certain nutrient-poor regions of Australia, ants disperse up to one-third of the plant species. These ant-dispersed seeds, however, don't seem to land in nutrient-rich soil any more frequently than seeds dispersed by other methods. The reason might be that Australian ants move around more and thus accumulate less garbage than their North American counterparts.

What, then, are the advantages of ant dispersal for Australian plants? Dunn pointed out that for millions of years native vegetation in Australia has burned as often as every seven years. Seeds buried underground by ants might gain a safe haven from seed-eating animals, pathogens, and desiccation until the fire's heat or some other aspect of fire stimulates them to germinate. Even though being carried by ants might be like trying to get out of town on the local city bus, at least the seeds end up underground.

SEEDS WRAPPED IN brightly colored fleshy fruits attract birds, mice, raccoons, coyotes, bats, primates, and other animals that eat the fruit, then drop or defecate viable seeds—often at some distance from the parent plant and ideally on fertile soil. Some animals you might not expect are seed-dispersing frugivores.

If you were out fishing in North America or another region in the temperate zone, would you tap the water surface with your fishing rod to lure fish? Probably not, if you wished to catch something. Most fishermen rely on stealth while tricking fish into taking bait. Not so in the flooded forests of the Amazon Basin, where fish compete to eat palm fruits, berries, and other fruits as they plop into the water. A "plop" made with the tip of the fishing pole draws in many hungry mouths.

Waters of these seasonally inundated flood plains along rivers are often low in dissolved minerals and nutrients, yet extraordinarily rich in fish species—including many that eat fruit. Some of these fishes crush seeds between their teeth. In doing so, they destroy the seeds. Others, such as various species of catfish, can't break the seeds' hard protective coverings. Experiments have shown that seeds pass through the catfishes' intestines unharmed. Whether fruits fall to the bottom substrate or get eaten by catfish, the seeds germinate once the forest is no longer flooded. Many kinds of seeds can stay intact in fishes' intestines for several days. Fish move around a lot, and some even undertake daily migrations of over twelve miles. Thus, catfish and other fishes may disperse seeds farther from the parent plant than if the fruit simply plopped into the water and landed on the muddy bottom.

Local fishermen in the Amazon Basin have long known that large numbers of fish enter the flooded forests when trees are fruiting. Time to get out the fishing poles. Once the fruits quit falling, the fish disappear.

Would you believe a fruit-eating frog? The only one known—a treefrog from Brazil—was reported in 1989. Frogs typically eat insects and other arthropods. Some frogs have been found with bits of plant material in their digestive tracts, but herpetologists generally assume incidental ingestion—the frog's sticky tongue snared not just an insect but also the leaf on which it was sitting.

The fruit-eating treefrog, which lives along the coast of southeastern Brazil, is different. Helio da Silva and his colleagues carried out their observations in an area of sand dunes between the ocean and a coastal lagoon. Vegetation in the habitat consists of shrub thickets and cacti. But there are also terrestrial bromeliads—plants in which rainwater collects in the clusters of long, narrow leaves. The frogs spend much of their time hunkered down in these moist microhabitats.

Discovery of the frogs' fruit-eating behavior was serendipitous—the

way many natural history observations are made. The investigators took
some of these treefrogs to their laboratory, and when the frogs defecated,
there were seeds in their feces. "What?" the scientists thought. "Frogs
don't eat fruit!" To figure out what was going on, they collected another
81 frogs at sunrise, presumably before the frogs had digested much of their
previous evening's dinners. A little over half the stomachs were empty,
but 29 percent of those with food had only small fruits and 13 percent
had a combination of fruits and arthropods. To learn whether the frogs
intentionally ingested fruits, the investigators offered fruits to frogs in
the laboratory. All the frogs ate the fruits. Two frogs fed nothing but these
fruits for four months survived fine.

Would seeds from the frogs' feces germinate? If not, the frogs would
be seed predators (destroyers) rather than seed dispersers. Another
experiment was in order. Amazingly (especially if you're a herpetologist
and you *know* that frogs don't eat fruit), the seeds extracted from frog feces
didn't croak. They germinated. The investigators suggested that because
these treefrogs defecate in the axils of moist bromeliads, the seeds might
have a better chance of germinating than those that simply dropped onto
dry sand.

The following year another Brazilian investigator, Roberto Fialho,
reported finding seeds of a shrub naturally germinating in the treefrogs'
bromeliads. Amazon lava lizards, though, also live in the dune habitat
and eat the same fruits. Fialho wondered which animal was the more likely
to do good things for seeds. The bromeliads in which treefrogs defecated
generally contain rainwater year-round, whereas the lizards defecated on
open, sandy soil. Fialho simulated seed dispersal by the frogs and lizards
by placing seeds from ripe fruits in bromeliads and on the soil next to
the shrubs. After 30 days he found that germination success of seeds left
on the sand was only 29 percent, as compared to 96 percent for seeds
left in bromeliads. Following germination in the terrestrial bromeliads,
the seedlings probably establish themselves after their roots reach
the soil.

So, let's all cheer for the world's only known fruit-eating frog that
likely helps out the plant by dispersing its seeds. If one treefrog does this,
there could well be others among the many hundreds of frog species about
whose habits we know very little.

Amazon lava lizards lost out in the contest with fruit-eating treefrogs,
but other lizards may be good seed dispersers. John Iverson found that
rock iguanas on the Bahamas and the Turks and Caicos Islands commonly
eat sea grapes and "seven-year apples." He studied the effect of seed pas-

sage through the lizards' guts by removing seeds of these two plants from fresh scats (yes, we field biologists are a rare breed). He planted the seeds, watered, and waited. Three months later, 38 percent of the sea grape seeds and 71 percent of the seven-year apple seeds had germinated. So, rock iguanas can disperse viable seeds. Iverson suggested that herbivorous lizards might be especially important seed dispersers on islands where they are the dominant fruit-eating vertebrates.

Some turtles and tortoises also disperse seeds. Omnivorous box turtles eat both fruits and animals. Hong Liu and his colleagues collected 145 Florida box turtles on Big Pine Key, in the lower Florida Keys. They placed their captives in plastic buckets containing two inches of water overnight—the water got the turtles' bowels moving. Ninety-five percent of the turtles' feces contained seeds of fleshy fruits. But do box turtles pass viable seeds? Liu and his colleagues planted the seeds of nine of the eleven fleshy-fruit species recovered from the turtles' feces. Germination percentage ranged from 10 to 79 percent.

The biologists then compared three groups of seeds from the plant with 79 percent germination success, a palm. One group was seeds recovered from turtle feces. The other two groups were seeds collected directly from plants: half with pulp left on, half with pulp removed. Compared to the 79 percent germination success of seeds recovered from turtle feces, only 28 percent of control seeds with pulp left on, and 39 percent of control seeds without pulp germinated. Seeds that passed through the turtles also germinated much faster than the control seeds with and without pulp. Three cheers for box turtles!

SOME BIRDS AND MAMMALS bury more seeds than they eat. Many of these stored seeds germinate.

Most conifers—trees and shrubs that bear their seeds in cones—rely on the wind to disperse their winged seeds after those fall from the cones. Until fully formed, the seeds inside the cones lie protected from seed predators such as squirrels and jays. Some pine trees, however, have the opposite relationship with seed-eaters. Piñon pines and certain other pines from mountainous regions of the western United States depend on seed-eating birds to disperse their wingless seeds. The cones often sit conspicuously on branches. As the cones open, some animals can see the seeds from twelve feet away. Many birds and mammals, including people, find these seeds irresistible. Why would piñon pines advertise their delicious, dark chestnut-brown seeds? Stephen Vander Wall and Russell Balda discovered that several species of seed-eating birds harvest many more piñon seeds than they eat.

Clark's nutcrackers, a type of jay, are the most efficient of these harvesters for two reasons. First, their long, sharply pointed bills make great chisels to pry open unripe cones, and their bills work as forceps to remove seeds from mature cones. Second, the birds have a pouch, unique to nutcrackers, that sits in front of and below the tongue. After packing up to 90 piñon seeds into its pouch, a Clark's nutcracker flies up to fourteen miles away and buries the seeds. The bird jabs its bill into the soil and makes a small hole. It disgorges up to fourteen seeds into the hole and then rakes soil or leaf litter over it with its bill. As a final gesture, the bird places a twig, pine cone, pebble, or some other small object on top.

Each Clark's nutcracker is worth several Johnny Appleseeds. Vander Wall and Balda found that during years of abundant pinecone production, nutcrackers bury and store seeds for about 50 days. If the cache site is three miles from the collecting site, a bird can make ten round trips per day. Using a mean of 65 seeds carried per trip, the investigators calculated that one nutcracker could store as many as 32,500 seeds each fall.

Fortunately for piñon pines, nutcrackers don't retrieve all their buried seeds. For whatever reason, these birds have an insatiable drive to store seeds, often three times more than they could possibly eat. Clusters of pine seedlings suggest evidence of forgotten seeds buried by nutcrackers. By dispersing seeds far from the parent plant—and planting them as well—nutcrackers help the parent trees colonize new areas.

HENRY DAVID THOREAU WROTE, "Though I do not believe that a plant will spring up where no seed has been, I have great faith in a seed. Convince me that you have a seed there, and I am prepared to expect wonders." Thoreau's faith in seeds mirrors our own. How we love to see sprouting seedlings in the spring—signs of new life. The next time you see green shoots, ask yourself what moved those seeds there. Ants? Birds or mammals? Maybe a reptile . . . or a frog!

GREEN, GREEN PLANTS OF HOME

Ballad of the Boll Weevil

Oh, the boll weevil is a little black bug,
Come from Mexico, they say,
Come all the way from Texas,
Just a-lookin' for a place to stay,
Just a-lookin' for a home,
Just a-lookin' for a home.

TRADITIONAL U.S. FOLK SONG

For over 100 years, cotton farmers in the United States have fought against boll weevils because of the homes these weevils choose. With their elongated snouts, these ¼-inch-long beetles pierce cotton flower buds

called "squares" and slurp pollen. They also puncture the immature fruits called "bolls" and eat the insides. Worst of all, they deposit their eggs deep inside squares and bolls. After hatching, the weevil larvae eat the insides of the cotton plant's reproductive structures. Damaged squares never develop into bolls, and damaged bolls don't produce fluffy fibers—the whole reason cotton farmers grow cotton.

Fig wasps are another of the many insects that live inside plants' reproductive structures. Nine hundred or so species of figs grow throughout tropical areas of the world. Odd as it might seem, the plants' hundreds of tiny flowers grow on the inside of the globular fruits. Each fig species has its own species of petite wasp pollinator.

Daniel Janzen described a fig-wasp association from Costa Rica as follows: A female wasp lands on an immature fruit and pushes her way in, breaking open a tiny plug at one end. In the process, she loses her wings. Inside, she moves from flower to flower and probes her ovipositor (egg-laying organ) down each style. If the ovipositor contacts the ovary, she lays an egg. While doing so, she also touches the flowers' stigmas and smears them with any pollen she may be carrying in two small pockets in her thorax. After laying all her eggs, she dies inside the fig. Sometimes only one female wasp enters a fig, but most often several enter the same fig and pollinate the flowers while laying their eggs.

The wasp larvae live inside seed coats and eat the developing seeds, though fortunately for the fig, not all the seeds—just those developing inside the ovaries penetrated by wasps' ovipositors. Loss of these seeds is the small price the fig "pays" to get extraordinarily effective cross-pollination. After about a month, the tiny wasps emerge from their seed coats. Wingless males emerge before females. They cut holes into the sides of the seed coats inhabited by females, insert their extensible abdomens, and mate with the confined females. Later, the newly mated females emerge from their seed coats and pack pollen from the fig's anthers into their pockets. Meanwhile, males cut an exit tunnel through the fig's wall for the females. The males don't use the tunnel—they die still trapped inside the fig. The winged females squeeze out, fly off to a tree with immature green figs, and begin the cycle anew.

Fig wasps and boll weevils aren't the only insects whose homes are seeds or fruits. During the 1950s my friends and I had contests to see whose "pet" Mexican jumping beans would jerk and tumble the longest. Little did we realize that, at least in part, our beans' performances reflected the temperature of our hands. As I recall, my beans were never stellar athletes. I must have had cold hands even back then.

Although Mexican jumping beans resemble small tan or brown beans, they are actually the separate sections of seed capsules from the jumping bean shrub. During the spring and early summer, female jumping bean moths lay their eggs on the ovaries of the shrubs' female flowers. After hatching, each yellow-white larva bores inside its new home. It eats the developing seed, creating a hollow for itself, and attaches to the inside walls with silken threads. By late summer, the capsule separates into three sections, or carpels. Each splits open and ejects its seed. Carpels with moth larvae, however, don't split open. Instead, they fall to the ground and tumble about for the next several months.

As approaching winter brings cooler temperatures, the larva uses its powerful mandibles to cut a circular opening partway through the carpel wall—its future exit door. Then it spins a silken cocoon around itself and remains motionless throughout the winter. The following spring or early summer the pupa pushes through its exit door, and a small gray moth breaks out from the pupal case. The moth mates, and if a female she lays her eggs. Both male and female moths die within a few days.

How does a moth larva "jump"? When warm, it grasps the silken threads with its forelegs, draws back its head and front part of its body, and bangs its head against the carpel wall. This snap transfers the force of the movement to the convex carpel and causes it to twitch, jerk, or tumble. Why does it do this? One speculation is that by moving about when exposed to hot sun, it might wriggle into cooler spots, perhaps into a crevice or shade. Other factors must be involved as well, however, because the larvae also move about while in the shade.

Called *brincadores* ("jumpers") in Mexico, the "beans" most often sold come from jumping bean shrubs in desert areas of the Mexican states of Sonora, Sinaloa, and Chihuahua. Lesser known is the Arizona jumping bean shrub, found in Arizona and certain desert areas of Mexico. In case you want to purchase some jumpers, you'll find them readily available online. Check out www.jumpingbeansrus.com or any of the other numerous sites that sell these engaging creatures.

PITCHER PLANTS HOLD water in "tanks" that provide ideal homes for certain aquatic or semi-aquatic animals. The elongate, tubular leaves of New World purple pitcher plants trap small pools of rainwater. Nectaries (nectar-producing glands) at the hood's base attract flies and other flying insects, and the lip opposite the hood serves as a convenient landing pad. But once an insect leaves the landing pad to try to reach the nectar, it loses its footing on the downward-pointing hairs sprouting from the hood. The insect falls into the water below. As in an insect nightmare, it can't climb back up the slick and waxy pitcher walls. It drowns, decomposes, and eventually nourishes the plant. How's that for a role reversal? Not "insect eats plant," but rather "plant eats insect."

Not all insects in pitcher plants are so unfortunate as those drowning flies. For the pitcher plant mosquito, the pitcher provides the larvae a sheltered nursery free of predators and rich with food. As the pitcher plant's victims drown and decompose in the water, the products of their decaying bodies nourish these mosquito larvae, which also eat bacteria and protozoans from this nutritious soup. The relationship is mutually beneficial. Pitcher leaves release oxygen into the water as they photosynthesize. As they breathe, the mosquito larvae give off carbon dioxide the plant can use in photosynthesis.

William Bradshaw and Christina Holzapfel found that female mosquitoes in Florida's Apalachicola National Forest lay their eggs in the youngest pitcher of a given plant. Appropriate choice of an egg-laying site is critical because the predatory effectiveness of the plants' leaves peaks between two and four weeks of age. A female mosquito's egg-laying behavior seems to anticipate the future nutritional needs of her offspring. Bradshaw and Holzapfel also point out that pitchers provide the mosquitoes with a haven from cold temperatures and thus "this mosquito—alone among the 300 most closely related tropical species—has been able to invade, and thrive in, a northern, temperate climate because of the protected habitat in which it lives: the stomach of a carnivorous plant."

Studies of pitcher plants from Borneo reveal that 33 species of invertebrates—including mites and the larvae of mosquitoes, hoverflies, and midges—live inside the six-inch jug-shaped pitchers of one particular species. This pitcher plant also shares a mutualistic relationship with ants that live in the swollen, hollow tendrils connecting the base of the pitchers to the leaves. The ants not only get a home, but they also feed from large nectar glands housed in two-inch-long fang-like thorns projecting over the "mouth" of the pitcher plant. In addition, the ants haul large prey, such as cockroaches, out of the liquid and up the pitcher wall, where they consume the carcasses. Removal of large prey might help the plant by preventing bacterial overload and putrefaction. Being sloppy eaters, the ants drop fragments back into the liquid. Thus, the plant doesn't lose out entirely on its cockroach catch.

A species of crab spider that lives only in these pitcher plants parasitizes the system. The spiders dangle from silken safety lines on the pitcher's inner wall. When insects enter to slurp from the nectar glands, the spiders intercept and eat them. The thieves also drop into the pitcher's liquid   still attached to their safety harnesses—and snarf up mosquito larvae.

Certain other plants also provide moist homes for animals. Many bromeliads, a large family of New World tropical and subtropical plants, have clusters of long, narrow leaves. Rainwater collects in the centers of the leaf clusters, creating homes for many insects and other aquatic or semi-aquatic animals. Most bromeliads are epiphytes ("air plants")—they grow on tree branches, but are not parasites. Instead, they get water and food from the air and from decaying organic matter near their roots. Some bromeliads, such as pineapples, grow on the ground.

Bromeliad crabs, small (less than an inch long) and reddish brown, live and breed in large ground-dwelling bromeliads in the Jamaican rain forest. Accumulated water in the bromeliads' leaf axils is low in calcium and dissolved oxygen—and acidic. It doesn't sound like a good nursery for baby crabs, and unmodified it isn't. The mother crab makes extensive home improvements during the nine weeks she cares for her young. She actively removes fallen leaves from the tank, increasing the water surface exposed to the air for gas exchange. She also dumps empty snail shells into the water. As the shells dissolve, they add calcium carbonate and buffer the water acidity. The shells also provide calcium for the young, a mineral critical for development and molting.

Tadpoles of certain species of poison dart frogs also develop in water-filled bromeliad tanks. Adults lay and fertilize eggs on moist ground, such as under leaf litter or inside curled leaves. Depending on the species, either the mother or father stays with the eggs and might urinate to moisten them. Once the tadpoles hatch, the parent hunkers down and the tadpoles slither up its back. They then ride piggyback to a water-filled bromeliad tank. The tadpoles develop in the water, where they feed on algae, detritus, and other organic matter.

In some species—for example, the strawberry poison dart frog from Costa Rica—the mother returns to feed her tadpoles after dropping them off at the nursery. First she carries her tadpoles, one by one, to several different bromeliad tanks. Later she stops by regularly and drops unfertilized eggs into the water, providing the tadpoles with the only food they eat until they metamorphose into froglets. Even more amazing, the tadpoles and their mother communicate with each other. The female peers into the bromeliad tank and then partially lowers herself into the water. If a tadpole is there, it swims close to the surface and repeatedly bumps its mother's hind legs. She lays an unfertilized egg. If no tadpole responds—perhaps it has died or already metamorphosed—she hops away without leaving an egg.

SOME ANIMALS MODIFY leaves to make their homes. Weaver ants from Africa, Asia, and Australia construct tent-like nests in the treetops. One large colony of weaver ants might extend through the crowns of three or more neighboring trees and consist of hundreds of leaf nests housing half a million or more ants.

The construction process begins when weaver ant workers grasp each other and form living bridges to pull two large leaves together. Meanwhile, others do the same, and eventually they maneuver leaves into a tent-like configuration. While these ants form rows and hold the leaves together, a second team of workers weaves the leaves together with strands of silk. The silk comes from their own ant larvae that are almost ready to pupate. The workers gently hold the larvae in their mandibles and use them as living shuttles. As the workers move the larvae's heads back and forth between two leaf edges, the young release silk threads from glands located beneath their mouths. They keep it up until they can produce no more.

Without silk, how do the larvae spin cocoons around themselves to pupate? They don't. Protected in the nest they've helped to make, they have bodyguards. If intruders approach the nest, workers boil out and defend the young.

Tent-making bats don't spin silk, but they also modify leaves for homes. Over a dozen species of bats in the New World tropics and at least two in the Old World roost in "tents" they make by cutting and folding leaves. The bats chew and sever the veins that branch out from the leaf's midrib, making the sides droop. They hook their feet into the leaves' undersides and hang there, concealed from above and from the sides. Each bat species makes its own characteristic type of tent, cutting the leaf in a particular way. Many prefer a certain size or shape of leaf, such as the elongated leaves of banana and bird-of-paradise plants, palm fronds, or the heart-shaped leaves of *Anthurium*.

The tents serve bats as roost sites; protect them from rain, wind, and sunlight; and shelter young while their mothers forage. Tents conceal bats from predators, but they also serve as an early warning system. Once an intruder jars the leaf, the alerted bats scatter.

Some species of bats make only one tent at a time, which they use until it falls apart—often lasting two months or more. Others make several tents and rotate their use, spending a few days in one before moving to another. Depending on the species and season, tents of socially monogamous species provide shelter for only one bat, a female and her baby, a mated pair, or a mated pair and its baby. Harem-forming species often make tents from large leaves that hold fifteen or more bats—a male, his entourage of females, and their gaggle of young.

FROM LIVING INSIDE plants' ovaries and "stomachs" to tenting with one's harem, many animals use plants for homes. Relationships between these animals and their plants range from mutually beneficial to destructive on the animal's part. Humans are included in the latter group—think worldwide deforestation for lumber to build our homes. But let's return to those boll weevils.

Boll weevils invaded the United States from Mexico in 1892. So much cotton was grown in the south that by 1922 the weevils had spread throughout the Cotton Belt, from western Texas to North Carolina. Within three or four years after the weevils reached an area, they often destroyed more than 80 percent of the local cotton crop. In desperation, many farmers switched to other crops, including peanuts; others migrated from the rural South to northern cities in search of a better life. What power this weevil has had over people!

A massive eradication project involving the insecticide malathion, begun in 1978, is working. By 2006 ten states had eradicated the weevils, and populations have declined in the other seven cotton-producing states. The U.S. Department of Agriculture expects the pest to be completely eradicated within the United States by 2009. So much for those boll weevils "just a-lookin' for a home" in U.S. cotton squares and bolls. Unfortunately, boll weevils are still expanding their range in South America . . .

POWERFUL PLANT PRODUCTS

A majority of the world's insects live in the Amazon rain forest, and the fact that the forest has not been devoured by this entomological onslaught is testament to these plants' abilities as chemical warriors. Plants protect themselves by producing an astonishing array of chemicals that are toxic to insects, thereby deterring them. When ingested by humans, these same plants—and their chemical weapons—may act in a variety of ways on the body: they may be nutritious, poisonous, or even hallucinogenic. And in some cases, they are therapeutic.

MARK J. PLOTKIN, *Tales of a Shaman's Apprentice*

Alkaloids, a class of chemical compounds, serve as major weapons for plants. Because these compounds taste bitter, they often deter plant-eating insects. People have long used the bitter taste as a clue that a plant has medicinal, stimulant, or hallucinogenic properties. In an odd twist, then, what deters insects attracts humans. Alkaloids include the stimulant caffeine; addictive substances in cocaine, heroin, and nicotine; the toxin in strychnine; the analgesic effects in codeine and morphine; and the hallucinogenic substances in mescaline and psilocybin.

Humans have learned how to use plants to heal and cure, control pain, stimulate or sedate the nervous system, increase endurance, season food, make themselves smell better, poison prey, seek pleasure or attain a state of ecstasy, heighten creativity, foresee the future, contact spirits and gods, and curse enemies with witchcraft. Some non-human animals also consume plants for reasons other than food. Animals, both human and non-human, develop strong relationships with certain plants. In most cases, the animal takes advantage of the plant's chemicals, often alkaloids. In all except one of the stories of powerful plants recounted here, chemicals impart the medicinal, stimulant, or mystical effects. But other properties of plants can be beneficial to animals also, as you'll see in the chimpanzee—nodular worm story.

HUMANS HAVE USED plants as medicines for thousands of years. For at least the past five millennia, the Chinese have treated colds, asthma, and bronchitis with *ma huang*, made from the evergreen plant *Ephedra*. The Ebers papyrus, written about 1500 B.C., recommended onion crushed in honey and consumed in beer to treat inflammation. Pliny the Elder, a first-century Roman scholar, extolled the virtues of garlic as an antidote against eczema, leprosy, toothache, asthma, hemorrhoids, and convulsions. Dioscorides, another Roman scholar, wrote about herbal medicines between 60 and 78 B.C. We might question its efficacy, but Dioscorides advocated wine made of wild cucumber to induce abortion. Galen, a Greek physician (A.D. 129–c. 199), prescribed opium from the opium poppy for headache, vertigo, asthma, epilepsy, and leprosy.

Many current conventional medicines come from plants. Morphine is derived from the opium poppy. Quinine, used to treat malaria, comes from the bark of the cinchona tree. Digitalis, used in treating heart disease, is made from the dried leaves of purple foxglove. The widely used anti-cancer drug Taxol comes from the Pacific yew tree. Ethnobotanist Mark Plotkin writes in his fascinating book *Tales of a Shaman's Apprentice:* "There exists no shortage of 'wonder drugs' waiting to be found in the rain forests, yet we in the industrialized world are woefully ignorant about the chemical—and, therefore, medicinal—potential of most tropical plants." He points out that few, if any, plants used as medicines—from contraceptives to cancer treatments—were discovered by trained botanists. Instead, the botanists learned of the plants' medicinal properties from indigenous peoples.

When it comes to non-human animals using plants as medicines, we have only to look in our backyards. Everyone who owns a dog or cat has no doubt watched it eat grass and later vomit. Grass also works at the other end of the intestinal tract, as a purgative. Whether grass serves as an emetic or a purgative, eating it helps shed intestinal parasites.

African chimpanzees host many parasites. One of the worst is the nodular worm, a parasitic nematode (roundworm). After being ingested accidentally by a chimp, the parasitic larvae penetrate the intestinal wall, cause inflammation, and form tumor-like nodules. Eventually juvenile worms leave the nodules, enter the inside of the intestine, feed on blood, and develop into sexually mature worms. Moderate infections cause diarrhea, weight loss, anemia, and abdominal pain, but severe infestations cause hemorrhagic cysts, septicemia (blood poisoning), and a blocked colon. Nodular worms are especially prevalent, and therefore most problematic for chimps, early in the rainy season.

For over three decades, scientists have watched chimps eat rough-textured leaves from trees that are not their normal food plants. The chimps ate these leaves during the rainy season. Some of the leaves were the hairiest leaves in the forest, a plant locally known as "African sandpaper." Most chimps seen swallowing these rough leaves were noticeably suffering from nodular worms, exhibiting diarrhea, abdominal pain, and malaise. Chimps folded the leaves concertina fashion, then held the leaves in their mouths for a few seconds before swallowing them whole. Because the leaves emerged in the chimps' feces undigested, it seemed unlikely that the chimps could extract useable chemicals that might fight intestinal parasites. In any event, the different species of these scratchy-leaved plants were chemically very different from each other.

Scientists eventually watched chimps across Africa swallow the leaves of nineteen different plants not commonly used as food. The common denominator of all the plants was rough-textured leaves. A close look at leaves from fresh feces unveiled the secret: live, wriggling nodular worms impaled on tiny barbs on the leaves' surfaces. The leaves were not killing worms by chemicals. They were removing worms via the "Velcro effect." By folding the leaves, the chimps increased the chance of hooking worms as they wriggled and got stuck in the folds. This clever technique of swallowing worm-hooking leaves has now been observed in at least eleven populations of common chimpanzees, bonobos (the smaller of the two species of chimpanzees), and eastern lowland gorillas in at least ten sites across Africa.

MOST OF US enjoy stimulants that crank our nervous systems into high gear, to give us a buzz. Stimulants come in many chemical forms and from many different plants. Although U.S. society has declared certain potent stimulants such as amphetamines illegal, mild ones such as caffeine found in tea leaves, maté leaves, coffee beans, and cacao beans (chocolate) form an integral part of social interactions. Caffeine wakes us in the morning and stimulates us into productivity throughout the day (and/or the night). The leaves of the coca plant contain the stimulant cocaine. Taken in its purified form, cocaine is highly addictive—and illegal, except for medicinal use. When the leaves are simply chewed, the chemical works as a mild stimulant, much like caffeine.

People reputedly discovered both coffee and coca by watching animals. According to legend, a goat herder in Ethiopia in about A.D. 600–800 noticed that after eating red berries (coffee beans) from a small shrub, his goats stayed awake all night, leaping about with exuberant energy. The goat herder ate some berries and felt so exhilarated that he shared his discovery with monks in a nearby monastery. The monks brewed a tea from the berries and were hooked because it kept them awake through long hours of prayer and meditation. They called their drink *hahveh*, meaning "stimulating and invigorating." Peruvian folklore tells that thousands of years ago people discovered the endurance properties of coca leaves by watching their pack animals—llamas. While carrying heavy loads for long distances, the llamas sought out and chewed coca leaves, which produced a sustaining effect. People tried chewing the leaves and gained the same benefit.

An estimated 15 million South Americans habitually chew coca leaves, grown mostly in Colombia, Bolivia and Peru. Although it is illegal to grow coca in Argentina, it is legal to possess and chew the leaves in the northwestern provinces. People openly sell and use coca leaves in the northern provinces of Jujuy and Salta, where, as in the rest of coca-chewing South America, there's an entire vocabulary surrounding its use. A *coquero* is a person who chews coca leaves. A *cocada* is the length of time a *bolito* (wad) of coca leaves sustains the *coquero*. The verb *coquear* means to chew coca. A *coquero* often keeps a supply of leaves in a leather pouch called a *bolsa para coca*. *Bica* is baking soda, chewed with the leaves to help extract the chemicals. Street vendors advertise their goods by yelling, "*Coca y bica.*"

While visiting Salta, my husband and I joined Argentine friends for a night of traditional folk music at an ancient mill, now a bohemian bar and restaurant where local musicians gather to jam. Everyone was chewing coca, from the waiters and patrons to the musicians, who sang with *bolitos* in their swollen cheeks. My husband and friends urged, "Marty, chew some coca. You'll need it to stay awake." At 1:00 a.m. I stuffed ten leaves into my mouth, took a little baking soda, and chewed. The leaves tasted a bit like horse manure smells. When we ordered empanadas a bit later, I expected everyone to discard his or her *bolito* into a napkin. Instead, they shifted their *bolitos* to one cheek and chewed on the other side. I followed suit. At 6:30 a.m. I was still wide awake and loving every minute of the music, the impromptu dancing, and the lively conversation. Could I have stayed awake without the coca? I don't know. I do know, though, that five minutes after hitting the pillow, I fell asleep. Once your brain says "sleep," coca doesn't keep you awake and jittery like coffee does.

PEOPLE THE WORLD over have used hallucinogenic plants, prized for their magical properties, for thousands of years. Many indigenous peoples used and still use hallucinogenic plants to gain access to the supernatural world and to speak to their gods. Once in direct contact with his gods, a shaman can diagnose disease and foresee the future through messages received back. People use hallucinogenic plants for pleasure. Certain plants induce an altered state of consciousness—changes in perception of reality, thought, and mood. The plants' chemicals transport us from reality into a dream world.

Indigenous peoples throughout the Amazon Basin brew a hallucinogenic drink called *ayahuasca* (pronounced ay-a-WA-ska), or "vine of the soul," from the bark of a woody vine. Warriors drink the brew before a hunt to become braver, improve vision, and foresee what animals they will kill. Shamans drink *ayahuasca* to diagnose an illness, divine its cure, and determine who caused the victim's disease. Participants drink *ayahuasca* during ceremonies, initiation rituals, and funerals.

If you were a 28-year-old conducting research in the upper Amazon Basin of Ecuador and were invited to partake in *ayahuasca*, would you try it? My host, Ildefonso Muñoz, invited me—and I did. The following excerpts from *In Search of the Golden Frog* describe my experiences after drinking the brew made by Quechua *brujos* (witch doctors).

Within about thirty minutes I felt dizzy. I knew the ayahuasca had taken effect when suddenly I was acutely aware of everything around me. Geometrical designs on the bamboo walls and floors wove in and out of focus. A large rat ran toward me. I didn't know if it was real or not. In a panicky voice I told Ilde that a rat was crawling over my foot. He shooed it away and I watched it run out the open door. Every time my eyes focused in a different direction, I saw my eyeballs jump as though they were attached to springs. Kaleidoscope patterns in brilliant colors danced about on the palm thatch above me. I closed my eyes and these images coalesced into a white golf ball falling down a long tunnel. Then the ball changed into a chicken egg, still falling down, until it became a round navel orange and disappeared. . . . Several frogs called back and forth to each other. I visualized the notes as a series of spiraling cadenzas inked onto an orchestral score. My eyes followed every note. A crumpled blanket by the door became a sheep's head with large eyes and buck teeth. I fearfully watched the head for movement.

. . . [T]hrough dense forest the orange, glowing eyes of a howler monkey pierced through me, perceiving my every thought. Slowly the woolly shape transformed into the grotesque head of a wrinkle-faced bat. Its black beady eyes were less threatening than the howler's fiery eyes. A hairy tarantula ambled across the bat's face, climbed into its left ear, and disappeared.

. . . I walked into a small thatched hut. A Quechua woman lay on a bamboo mat, moaning and groaning in the final stages of labor. I touched her sweaty forehead and then I became the woman. As my uterus contracted, I gasped with pain. I was alone. Then another contraction—more intense this time. I pushed and I groaned. Relief followed as I watched the baby emerge from between my legs.

A profound mind-altering experience, *ayahuasca* took over my mind and body, heightened my senses, and diminished the borders between reality and fantasy. But more than that, it was a learning experience. I became that Quechua woman and gave birth. Six years later, when my husband and a nurse-midwife helped me deliver my firstborn, without drugs, I felt as though my mind and body had been through childbirth before. Having experienced that, I can only imagine the power the plant holds for warriors and shamans. No wonder many consider it a plant of the gods.

The Tukano, a tribe of Amazonian Indians, claim that jaguars claw and gnaw at the bark of the *ayahuasca* vine and chew the leaves. Tukano Indians drink *ayahuasca* to enhance their sensory awareness and night vision. They believe jaguars gain the same benefit. But would jaguars really gnaw at a vine to see better? I wonder. Cats naturally have superb night vision! Perhaps jaguars see kaleidoscope patterns in brilliant colors dancing through the forest. Or hairy tarantulas crawling into bats' ears . . .

Peyote, a small spineless cactus that grows in northern Mexico and southern Texas, is another powerful hallucinogen. Chewing peyote buttons, cut from the dried crown of the cactus, allows the partaker to enter another world. Peyote cactus contains over 25 psychoactive alkaloids, the most powerful being mescaline.

Native Americans in Mexico have likely chewed peyote buttons for thousands of years, judging by its presence in ancient cliff dwellings and pottery relics. Kiowa and Comanche warriors returning from raiding parties in Mexico probably brought the peyote cult north around 1880. Use of peyote spread quickly throughout the Southwest and the Great Plains because of its unique qualities—and its power. The social anthropologist James Slotkin eloquently expresses the power of peyote use by Native Americans as follows: "The white man goes into his church house and talks about Jesus; the Indian goes into his teepee and talks to Jesus." Although the U.S. government tried to ban the drug, its use increased and became incorporated in religious ceremonies and worship, Christian and pagan. In 1962 members of the Native American Church won a test case against the state of California, which contested the religious significance of peyote. Five years later the U.S. Congress legalized peyote as a sacrament. Some tribes of Native Americans still widely use peyote for religious purposes.

Some non-human animals indulge in hallucinogenic plants. The strong smell and bitter taste of *Datura*—a poisonous and hallucinogenic plant with large, white tubular flowers—repulses most animals. But not hawk moths. After feeding on *Datura* nectar, hawk moths appear disoriented. Their flight is erratic, and they have trouble landing on flowers. Yet they keep coming back for more nectar. In Asia water buffalos working in fields of cultivated opium poppies graze on the flowers. Ingestion seems to be intentional, judging from the fact that the poppies' bitter taste warns: Toxic! The water buffalos don't eat enough to be poisoned—just enough to "feel good." By season's end, the water buffalos are restless and exhibit tremors and convulsions—signs of opium withdrawal.

WITHOUT PLANTS, there would be no animals as we know them on Earth. Plants provide us with food, clothing, shelter, and oxygen. Beyond those necessities, they offer inspiration through their beauty and improve our quality of life—whether fresh or fermented, healthful or hallucinogenic, medicinal or mystical, sacred or stimulating. What more could one ask of a plant than to hook worms, keep us awake, and transport us to other worlds?

THERE'S THE RUB

Much to his astonishment, Sigstedt found that when he gave the bear root to bears in the Cheyenne Mountain Zoo in Colorado Springs, they immediately began to chew up the root and rub it over their bodies—precisely what many Indian legends say that the bear taught humans to do.

REBECCA ANDREWS, "Western Science Learns from Native Culture"

During the 1970s, ethnobotanist Shawn Sigstedt learned from a Navajo family the legend that Bear gave Navajos the powerful medicinal plant *osha*, or bear root: *Ligusticum porteri*. After he heard similar accounts from other Native American tribes of the southwestern United States, he offered the root to captive bears in the Cheyenne Mountain Zoo—and watched them rub. Brown, American black, and Kodiak bears all chew *osha* and rub the saliva-laden paste through their fur. Many cultures consider bears powerful healers because they seek out plants to cure themselves, and in fact most Native American populations living near bear root use the plant to treat fungal and bacterial infections. This widespread member of the parsley and carrot family grows above 7,000 feet throughout the Rocky Mountains from Canada to Mexico. Mexicans commonly call the plant *chuchupate*, an ancient Aztec term meaning "bear medicine."

Birds and primates also rub plants onto themselves. In some cases, experiments document a medicinal benefit to the behavior. In other cases, we can only speculate on the reason.

OVER 200 SPECIES of birds rub ants or other arthropods onto their feathers, a behavior called "anting" (see "Cow Pie No. 5"). But birds don't always "ant" with arthropods. Some birds rub tobacco, mustard, onions, lemons, limes, and parts of other plants onto their feathers.

One July day Dale Clayton watched a male common grackle struggling to balance himself atop half a lime. Once stable, the bird hammered at the lime, then preened himself with bits of fruit held in his bill. Others have also reported seeing grackles "ant" with lemons and limes. These observations prompted Clayton and a colleague to carry out a simple experiment to see whether lime kills chewing lice, common grackle parasites that feed on feathers and reduce mating success and even survival of their hosts.

The investigators removed a primary feather and its attached lice from each wing of a live rock dove (easier to obtain than a grackle) and placed the feathers into two petri dishes. They put a lime slice into one petri dish and an equivalent amount of tissue paper soaked in distilled water into the other dish. The feathers touched neither the lime nor the tissue paper. Clayton and his colleague examined the lice under a microscope nine hours later. Thirty-five of 52 lice (67 percent) in the dish with the lime slice had died; most of the other 17 were immobile except for trembling legs, perhaps in their death throes. In contrast, only 1 of 31 lice from the dish with water-soaked tissue paper had died; one appeared to be dying, and the other 29 looked chipper. Although the experiment was not replicated, the difference was intriguing.

Clayton and his colleague wondered if the apparent insecticide was in the lime's juice or its peel. Using a fine-tipped brush, they dabbed lime juice onto the heads of 7 lice. After twelve hours, the lice were still alive and apparently healthy. Thinking that maybe more juice would do the trick, they drenched 9 lice with lime juice. Still no effect. Next they dabbed lime peel extract onto the heads of 10 lice. Within seconds all died, suggesting that the lice in the original experiment died from lime-peel vapor, since they had not come into physical contact with the lime slice. Other studies have shown that citrus oils contain substances toxic to fleas on cats and guinea pigs. So, common grackles may indeed delouse themselves with a lime-peel rubdown.

WHITE-FACED CAPUCHIN MONKEYS spend much of the day traveling through the trees in troops, screaming, whistling, yipping, and barking as they forage for ripe fruits and arthropods. Mary Baker studied four troops of white-faced capuchins in Costa Rica. She watched monkeys rub plants of at least four different genera onto their fur, most commonly several species of citrus:

The application of plant material was highly energetic, almost frenzied in appearance. The monkeys moved rapidly, drooling, biting into the plant and rolling it between their hands. Then, using their hands, feet, and tails, they applied the plant material over various body parts or over their entire bodies. When using citrus fruit, sometimes only the rind was used. They first abraded it by biting or rolling and pounding the fruit on tree branches and then applied it over the body. More commonly, however, after pounding the fruit on tree branches, the monkeys broke open the fruit to extract the pulp and juice. The monkeys hugged the fruit to their chests and stomachs while simultaneously rubbing it, digging into it, and rubbing it into their fur, usually covering the entire body.

Sometimes individual capuchins rubbed plants into their fur, and sometimes the entire troop got into the act. Baker describes the scene as "a mass of wet, drooling monkeys with bits of citrus pulp and juice or broken leaves sticking to their fur, squirming and rolling over and around each other."

Baker found that the monkeys rubbed their fur with citrus more often during the wet season than the dry season. The warmer temperatures and higher humidity of the wet season presumably intensify bacterial and fungal infections, and of course mosquitoes are more abundant. Lacking experimental data to document a medicinal benefit of citrus for the capuchins, however, Baker pointed out that the monkeys might take citrus baths for other reasons as well. Rubbing citrus onto fur might establish a group scent. Social fur rubbing might reinforce social ties, much like grooming behavior. Or the monkeys might simply enjoy the pleasurable or stimulating feeling of citrus on their skin.

After reading Baker's paper, I remembered that decades ago several Ecuadorian colleagues swore by lemon peel as a mosquito repellent. Their advice to me: "Just peel a lemon and rub the inner rind over your face and arms." I tried it. Sure enough—we smelled like freshly squeezed lemonade while frog-hunting in the rain forest, but it repelled the blood-thirsty mosquitoes. According to Baker, Latin Americans also use lemon as a poultice, antiseptic, and to soothe insect bites. (Of course, if you had given yourself a lemon-peel rubdown in the first place, you wouldn't have those itchy mosquito bites.) In humans, lime is an effective disinfectant for newborns' eyes, hemostat for nosebleeds, mosquito repellent, and relief for itching mosquito bites. Sweet orange and mandarin orange make good bactericides. The latter is also a fungistatic, a substance that inhibits growth of fungi without destroying them. Sour orange serves as a poultice for sores and a fungistatic. I like to think that capuchins use citrus rind as mosquito repellent, just as I did so many years ago.

CENTRAL AMERICAN SPIDER MONKEYS have long arms and legs and prehensile tails. Resembling gangly hyperactive spiders, the monkeys swing through the treetops by their arms. When not traveling, they often sprawl along branches, now looking more like drowned daddy longlegs. Like capuchins, spider monkeys self-anoint with plants, but with leaves instead of fruits.

Mattias Laska and two colleagues studied a group of spider monkeys in Veracruz, Mexico: six adult males, three adult females, and one infant. They watched two males rub three different plants onto their fur: Alamos pea tree, trumpet tree, and wild celery, each of which has leaves that release an intense, aromatic odor when crushed. The males bit into or chewed leaves, then rubbed saliva and plant fragments into their armpits and onto their breastbone areas for 30 seconds to two minutes.

Why the peculiar behavior? The investigators ruled out insect repellent or antibiotic as the benefit for the following reasons. Anointing behavior did not correlate with time of day, season, humidity, or temperature. The other seven adults did not rub leaves over themselves. The two males anointed only limited areas of their bodies. Surely mosquitoes, ticks, lice, bacteria, and fungi go for additional body spots—not just armpits and breastbone areas. Finally, the plants aren't particularly effective against insects or pathogens, though they yield mild benefits.

Laska and colleagues speculated that male spider monkeys might self-anoint to communicate social status or increase sexual attractiveness. Just as some human males spritz cologne, some male spider monkeys might use aromatic leaves to improve their body odor. One of the two males that self-anointed was the highest-ranking monkey in the group and the father of the only infant. Spider monkeys do have a highly developed sense of smell for fragrant plants. Exactly *what* the monkeys might communicate with aromatic plants we don't know, but they could be using plants as perfumes.

NATIVE AMERICANS LEARNED the medicinal properties of *osha* by watching bears rub it onto their bodies. Later, pioneers to the American West learned of its value. One common name for this plant is Colorado cough root, because early pioneers to Colorado used it to treat coughs. Scientists have identified over a dozen compounds of known pharmacological activity in bear root. You don't need to search for the plant, then dig up the roots. You can buy *osha* in health food stores and on the web—fresh or dried root, liquid extract, or capsules. *Osha* is recommended for treating more than bacterial and fungal skin infections. Suggested ailments for which *osha* is recommended include flu, colds, sore throat, allergies, asthma, indigestion, and insect repellent. My husband, Pete, drinks *osha* tea for the laryngitis he often gets from talking too much.

People also might have learned long ago from monkeys and birds that citrus fruits repel insects and kill pathogenic fungi and bacteria. Products containing the citrus peel oil extract limonene can control fleas and ticks on your pets and kill ants, fleas, roaches, and silverfish in your home. For yourself, there's Repel Lemon Eucalyptus, effective against mosquitoes, deer ticks, and no-see-ums (sand flies) for up to six hours. That sure beats rubbing lemon rind on your face and arms!

And what about our colognes and perfumes? Did we learn how to make ourselves smell more attractive to the opposite sex by watching animals rub their fur with plants? People have made perfumes from fragrant plant oils for thousands of years. The earliest culture to do so may have been the Egyptians, who even took their perfumes with them to the tomb. Even before perfumes were "invented," though, Egyptians burned incense. Perhaps one day an Egyptian courtesan watched her pet monkey rub an aromatic plant onto its fur and shouted, "Aha!" Instead of smelling good by burning incense, she could rub plant oils directly onto her body and attract a wealthy suitor.

ANTS AND PLANTS

I had the common red passion-flower growing over the front of my verandah, where it was continually under my notice. It had honey-secreting glands on its young leaves and on the sepals of the flower-buds. For two years I noticed that the glands were constantly attended by a small ant (*Pheidole*), and, night and day, every young leaf and every flower-bud had a few on them. They did not sting, but attacked and bit my finger when I touched the plant. I have no doubt that the primary object of these honey-glands is to attract the ants, and keep them about the most tender and vulnerable parts of the plant, to prevent them being injured. . . .

THOMAS BELT, *The Naturalist in Nicaragua*

The ants place large quantities of insect parts in the plants, which probably take up their decomposition products.

DANIEL H. JANZEN, "Epiphytic Myrmecophytes in Sarawak"

In 1874 Thomas Belt proposed that certain plants feed certain ants. One hundred years later, Janzen published his complementary speculation that certain ants feed certain plants. And how right they both were! Organisms—plants or animals—that spend at least part of their life cycle with ants are called "myrmecophiles," from the Greek *myrmex,* meaning "ant," and *philos,* meaning "loving." Let's look first at some myrmecophilous plants that feed ants.

"OW!" I HAD brushed my hand against a bull's horn acacia tree in a lowland Costa Rican dry forest—and gotten stung by angry ants protecting their tree . . . just as Thomas Belt had gotten bitten by ants guarding his red passionflowers. Elsewhere in his book, Belt wrote of "standing armies" of ants kept by acacia trees in Nicaragua. This ant-plant relationship is one of classic mutualism. Bull's horn acacia trees, named for the impressive pairs of thick thorns distributed along the branches, rely on ant bodyguards to protect them from herbivores. The ants live inside the hollow thorns and depend on the acacias for shelter.

Many ant-loving plants bribe plant-loving ants with food to keep them around. Some myrmecophiles, including red passionflowers and bull's horn acacias, produce nectar in exposed glands called "extrafloral nectaries." These tiny cup-like structures sometimes occur on the outside of flowers, as well as on leaves, stems, or other plant parts. Extrafloral nectaries have nothing to do with flower pollination, although butterflies and even hummingbirds sometimes stop by for a quick sip. Instead, the glands attract ants with a sweet nectar of sugars and amino acids. Worker ants patrol their home plants and defend their nectaries. They capture or dislodge plant-eating insects and frighten away female butterflies seeking to lay eggs. In doing so, they protect the plant from herbivores. Fewer leaves munched increases the plant's reproduction and/or chances for survival. Some plants, including the bull's horn acacia, offer their bodyguards an additional reward—little pale, waxy nutritive buttons on their leaf tips called "Beltian bodies," named after Thomas Belt. These food bodies are rich in proteins, lipids, and carbohydrates—great for both adult and larval ants.

I can't leave acacias and their Beltian bodies without sharing an old OTS (Organization for Tropical Studies) joke from Costa Rica: How can you spot a bull's horn acacia in the dark? From the sounds. What sounds? The Beltian (belchin') bodies!

Black cherry trees are classic myrmecophiles, and their ant bodyguards are fierce—especially against hungry caterpillars. But first a brief description of the three players: tree, ant, and herbivore. When black cherry trees first leaf out in early spring, each leaf offers 60 to 90 active extrafloral nectaries—bribes for ants. A week later, only the 2 to 5 nectaries at the leaf base actively produce nectar. By three weeks, most leaves have no active nectaries. Red-headed ants in these cherry trees eat insects, including leaf-munching caterpillars, but they also love the dessert provided for them by the black cherry trees. An ant spends about 55 seconds slurping nectar from a leaf, then about 13 seconds moving to an adjacent leaf. The 20,000 or so worker ants from a colony forage on black cherry trees within about 60 feet of their nest. The herbivores—eastern tent caterpillars—are striking: dark gray with a yellow racing stripe down the center of the back and matching yellow, blue, or white spots along the sides. Wispy cream-colored or beige hairs cover their bodies. These herbivores forage as family groups in black cherry trees, devouring the leaves of one branch and then moving to another. A tree with several such family groups may be completely de-foliated by very hungry caterpillars. Caterpillars in the final stages—think voracious teenagers—cause most of the damage.

David Tilman studied black cherry tree/tent caterpillar/ant interactions in a field in Michigan during the 1970s. He watched ants grab young caterpillars in their mandibles and haul them back to their nest. Early in the season, when the ants were much larger than the caterpillars, the ants easily captured and transported their prey. As the tent caterpillars grew to ant size, though, the ants struggled to conquer the ravenous larvae. In fact, ants successfully attacked caterpillars only during the first three weeks following a tree's leafing out—until the time when caterpillars were about twice ant size. As the larvae grew larger, the ants waved their antennae at them but did not attack.

Tilman found a difference in leaf damage between black cherry trees close to ant colonies and those farther from the colonies. As you've already guessed, the trees farthest from ant colonies had the most late-stage caterpillars crawling over the branches and leaves, and those trees suffered the greatest defoliation. In contrast, the ants did a good job of protecting trees within their foraging range from very hungry caterpillars. On trees nearer to colonies, extrafloral nectaries attracted ants early on—recall that to begin with, each leaf offers 60 to 90 sweet bribes. The ants devoured many small caterpillars before the latter got too big for the ants to handle. The result was less damage to the trees' leaves. Recall also that by one week, only the 2 to 5 nectaries at the leaf base actively produced nectar, and that by three weeks most leaves had no active nectaries. It turns out that the cherry trees secreted nectar and attracted ants only during the time when the caterpillars were small enough that the ants could capture them. Thomas Belt would have appreciated Tilman's observations. I can imagine him exclaiming, "How prudent! Why waste energy and nutrients feeding ants when they can no longer be of any help?"

Back to the tropics and a clever experiment. The Brazilian tree *Lafoensia pacari* produces extrafloral nectaries on its new leaves. Ana Korndorfer and Kleber Del-Claro removed all ants from 17 trees that had new leaf buds and applied a gummy substance called Tanglefoot to the base of each tree so ants couldn't climb up. They left ants on another 23 trees with new leaf buds—no Tanglefoot on these trees. The extrafloral nectaries produced nectar until the leaves fully expanded, and on trees without Tanglefoot ants visited these glands throughout most of the first three months of the experiment. After three months, when the glands on the mature leaves no longer produced nectar and few ants visited those trees, Korndorfer and Del-Claro examined leaves from each of the 40 trees. They found that the trees without ants had significantly more damage from herbivores than trees with ants. Not surprising, but . . .

The story didn't end after three months—or with the ants. This Brazilian tree and many other plants accumulate silicon in their leaves, which toughens the leaves and deters herbivores. Six months after beginning the experiment, Korndorfer and Del-Claro analyzed the silicon content of leaves from each of the 40 trees. Trees with bodyguard ants had accumulated significantly less silicon in their leaves than had trees lacking ants. By the end of six months, the two groups of trees no longer differed significantly in herbivore damage to their leaves. One group had benefited from ant protection, the other from silicon protection. Apparently, if ants aren't available, these trees invest in an anti-herbivore backup plan: chemical defense. The trees are switch-hitters!

AND NOW THE flip side: Some ants feed plants. Myrmecophytes, or "ant plants," have specialized hollow structures that shelter ants. In return for free housing, ant residents bring in food, a behavior called "myrmeco-trophy" (from the Greek *trophia*, meaning "to nourish").

In 1974 Daniel Janzen, who had earlier done the pioneering work with acacia trees and their ant bodyguards, reported on four species of myrmecophytes from Sarawak, Malaysia. These myrmecophytes were epiphytes—plants that, like many orchids and bromeliads, grow on other plants but make their own food. Virtually all individuals of these four species of epiphytes at Janzen's study site had tiny non-aggressive golden ants (*Iridomyrmex cordatus*) living in them. The plants provided relatively dry homes for golden ants in an otherwise rain-soaked environment. Great for the ants, but did the plants gain anything from these petite, passive occupants? Janzen noticed something peculiar about these ants. Most arboreal ants throw refuse outside the nest entrance, but these golden ants stuffed their garbage—worker ant and termite heads, mites, beetle heads and wings, heads of parasitic wasps, spider legs and carcasses, centipede segments, and occasional seeds—inside their living nests. Janzen suspected that golden ants "feed" their plant hosts rather than protect them from herbivores. How? He proposed that the plants might absorb the decomposition products of the arthropod fragments. He also speculated that because these four epiphytes live on vegetation growing in nutrient-poor soils, they depend on the ant-imported nutrients for their survival.

Two of the four epiphytes Janzen studied belong to the coffee family Rubiaceae. Both have a single large tuber (swelling) at the base of the stem. *Hydnophytum formicarium* is called "truffle plant," in reference to its smooth truffle-shaped tuber. *Myrmecodia tuberosa*, called "ant house plant," has a spiny tuber. Ants nest in cavities of both plants' tubers. Janzen noticed that parts of the tuber cavities have smooth walls; other areas are rough. He found that worker ants leave arthropod fragments and nestmate carcasses in rough-walled areas lined with absorptive tissue, and they sequester their brood in smooth-walled areas lined with tough, impervious tissue.

A third epiphyte species, the "ant fern" *Phymatodes sinuosa*, consists mostly of swollen, hollow rhizomes (thick, root-like horizontal stems), which form a dense mat over a tree branch. Ants enter through broken ends of old, rotting rhizomes. The ants first pack discarded prey remains into the ends of a cavity and then, eventually, fill the center. They house their brood in the same cavity but separate the "kids" from the garbage.

In the fourth epiphyte species, the Malayan urn vine *Dischidia rafflesiana*, ants nest in clusters of rolled-up "ant-leaves." The ants first use a leaf cavity as shelter for their brood, later for both the brood and refuse, and finally only for refuse.

Janzen had no proof that ants fed the plants, but his observations led to the reasonable hypothesis that plant tissues took up ammonia from the ants' rotting garbage. The idea needed to be tested. Janzen ended his paper by suggesting: "Two classes of experiments are badly needed. The physiologist needs to feed radioactively labeled insects to the ants to see if this material ends up in the plants. . . . The other experiment, and in my opinion the more important, is to compare the seed production rates (or failing in that, the growth rates) of plants with and without their ants."

In the late 1970s, investigators tested Janzen's hypothesis that ants feed truffle plants and ant house plants, the two epiphytes with absorptive tissues in the cavities of their tubers. In one experiment, investigators gave ants food tagged with radioactive tracers. The tracers ended up in the rough-walled chambers of the ant house plant, presumably through the ants' defecation. In another study, products from radioactively labeled fruit fly larvae left in the plants as ant garbage ended up in the tissues of the truffle plants. These studies documented that the plants do indeed absorb nutrients from the ants' wastes. Just because the plants absorb nutrients, though, doesn't mean they can metabolize them. Do the plants gain a reproductive or growth advantage? There is almost certainly some benefit, but to my knowledge no one yet has followed up on Janzen's suggestion to look at seed production or growth rate in these specific epiphytes.

Another ant-fed epiphytic myrmecophyte is the cow horn orchid, found in nutrient-poor habitats from southern Mexico through most of Central America. A cow horn orchid has up to 40 hollow pseudobulbs, aerial

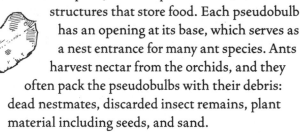

structures that store food. Each pseudobulb has an opening at its base, which serves as a nest entrance for many ant species. Ants harvest nectar from the orchids, and they often pack the pseudobulbs with their debris: dead nestmates, discarded insect remains, plant material including seeds, and sand.

In the late 1980s, Victor Rico-Gray and colleagues experimented to see if cow horn orchids actually absorb organic material from the ants' garbage dumps.

They fed fire ants radioactively labeled honey, then killed the ants and packed them into pseudobulbs. One to eight weeks later they examined plant tissue for radioactivity. Their results echoed those of experiments with the Southeast Asian epiphytes. The radioactively labeled material moved from the ants into the plants, including their actively growing roots. The longer the pseudobulbs held the "hot ants," the more the material was incorporated into orchid root, stem, and leaf tissue.

And now we get a step closer to finding out what's going on with myrmecotrophy. In 1995 Kathleen Treseder and her colleagues tackled the question: Do epiphytes that absorb nutrients from ants actually use those nutrients? They focused on a Malaysian epiphyte, *Dischidia major*, that grows in nutrient-poor soil. Ants live in sac-like leaves of this epiphyte, where they both raise their young and deposit their garbage. By measuring stable isotope ratios of carbon and nitrogen of ants, plants, and substrates, the investigators tested the hypothesis that the plant absorbs ant-respired carbon dioxide, and that the plant uses ant debris as a source of nitrogen. *Voilà!* Their results revealed that 39 percent of the carbon in the leaves of ant-occupied plants comes from the resident ants' respiration, and that 29 percent of the nitrogen in the epiphytes comes from the ants' garbage. No doubt about it—these ants earn their keep. They provide significant amounts of carbon and nitrogen that the plants do indeed use in their otherwise nutrient-poor environment.

So far, all of these myrmecotrophy relationships have involved epiphytes. But ants feed other plants as well, and the plants also incorporate these nutrients into their tissues. For example, experiments with *Cecropia* trees in Trinidad reveal that over 90 percent of a tree's nitrogen comes from garbage deposited by ants that nest in the tree's hollow stems.

Another non-epiphytic myrmecotroph is *Maieta guianensis*, an understory shrub from French Guianan rain forest that has unusual leaves: two-chambered pouches sit at the bases of the leaf blades. *Pheidole minutula*, a "big-headed ant," uses one chamber as shelter and fills the other with its feces and prey remains. Experiments have shown that these shrubs receive about 80 percent of their nitrogen from the ants and their waste. The internal surfaces of the leaf pouches bear nipple-like protuberances, once thought to be organs that supply ants with food. (Intriguing image . . . an assumption that likely reflects our mammalian bias.) Further analysis, however, suggested that they are not secretory organs, and in fact transfer of nutrients is probably in the *other* direction—from ant to plant. Most likely the protuberances absorb nutrients from the ants' garbage dumps. Because this shrub grows in the shade on poor soil and competes with large trees for nutrients, ant waste might be critical for its survival.

We've learned a lot since Janzen suggested that ants feed plants. Nutrient absorption from piles of debris that ants leave around has been demonstrated or suggested for over 60 species of plants from ten families, with many ant genera involved. Experiments have shown that in certain plants, individuals associated with ant nests grow significantly larger than individuals not associated with ant nests. The ant-plant relationship in many mrymecophytes is complex, as ants not only feed the plants but also protect them against herbivores. In some of these plants, removing ants has led to increased herbivore damage and greatly reduced fruit production. Although these ants definitely do good things for the plants, sometimes it's hard to separate the beneficial effects of food versus protection.

TROPHIC RELATIONSHIPS BETWEEN plants and animals aren't always what we expect. Who would have thought that plants feed ants and ants feed plants? But Thomas Belt's proposal that plants' "honey glands" attract ant bodyguards, recorded over 130 years ago, and Daniel Janzen's speculation that some ants feed plants, published 100 years later, were right on. With our current emphasis on molecular level research, let's not forget about the value of good old-fashioned field observations.

The next time you reach down to pick a flower and get stung or bitten by an irate ant, think about what it might be doing for the plant—and what the plant might be doing for its bodyguard.

4 Invasion of the Body Snatchers, & Other Interactions with Fungi and Bacteria

DOES YOUR SKIN ever crawl, sensing another presence in the room? If so, it should be crawling all the time because you're never alone. You host trillions of silent, invisible bacterial and fungal guests. Although tiny, they play gargantuan roles—for better or worse—in your life. They do the same for other animals and for plants.

Consider bacteria, single-celled organisms. Bacteria are unique in their simple cell structure. They're prokaryotic, meaning "before the nucleus." A bacterial cell lacks a well-defined nucleus surrounded by a membrane, and it lacks other membrane-bound structures called "organelles" that characterize the cells of all other organisms, including our own. At least 90 trillion of "our" cells are actually bacteria, and they make up 10 percent of our weight! Fortunately for us, most of our bacteria are commensals— they share our food but do not harm us. Many help us just as we help them, and fewer still are parasites.

Bacteria are the oldest known life-forms on our planet. In fact, the oldest fossil bacteria, 3.5 billion years old, are virtually indistinguishable from some modern-day species. Bacteria are also the most abundant organisms on Earth. A teaspoon of fertile soil might house 2.5 billion individual bacteria. They live in just about every habitat—from hot springs, salt flats, and beneath Antarctic ice to plants' roots and animals' guts. They reside 4,500 feet below Earth's surface and at ocean depths of 30,000 feet. They live in the clouds and in the snow on the highest mountaintops. Most bacteria "eat" molecules of food that enter pores in their cell membranes.

In contrast to a bacterium, the cell of a fungus has a membrane surrounding the nucleus and each organelle, separating these structures from the jelly-like fluid called "cytoplasm." Yeasts and some other fungi are single cells, but most fungi are multicellular. Like bacteria, fungi live in and on us. Most live in ecological balance with our other microorganisms. Some fungi help us by keeping pathogens at bay. When things get out of whack, however, certain fungi can kill us.

Fungi live almost everywhere: in water, in soil, between your toes, on forgotten bagels in your bread box. Fungi lack chlorophyll and thus cannot make their own food. Unlike animals, fungi don't ingest and then digest other organisms. Instead, fungi excrete digestive enzymes into their environment and absorb nutrients from the resultant breakdown of the dead or living organisms around them. Specialized soil fungi provide water and nutrients to many land plants; in turn, the fungi receive carbohydrates the plant synthesizes. We often think of wood-eating fungi as passive organisms, but some also are predators. Besides working as the forest's garbage disposals, they attack and consume living organisms, including nematodes (roundworms) and amoebas. The edible oyster mushroom displays one of these "Jekyll and Hyde" personalities. We eat mushrooms and truffles, and we use yeasts to bake bread and brew beer. And then there are the parasitic species that "snatch bodies."

For those unfamiliar with fungi, a brief description of their anatomy is in order. Multicellular fungi are composed of filaments called "hyphae" (singular = hypha). An individual's hyphae are collectively called a "mycelium" (plural = mycelia). Most of a fungus's body, the mycelium, lives underground, inside a dead tree, or otherwise out of sight. Fungal bodies range in size from barely visible without a microscope to a fungus from eastern Oregon that is among the world's largest organisms. Its mycelium covers 2,200 acres of forest, weighs over 330,000 pounds, and is estimated to be 2,400 years old. Fruiting bodies—for example, mushrooms—are reproductive structures consisting of tightly packed hyphae that produce spores, asexual cells capable of developing into new organisms without uniting with other cells.

Bacteria and fungi, highly successful organisms that can multiply rapidly, are extremely adaptable and versatile. Bacteria can mutate quickly. Many bacteria and fungi remain dormant until just the right living or reproductive conditions come along. Some produce antibiotics—substances that inhibit other microorganisms that compete with them for space and other resources.

Examples of bacterial and fungal activities in the following essays reveal the profound influence these organisms exert on plants and animals—good, bad, and downright scary.

INTESTINAL MICROBES AND THE GAS WE PASS

Breaking wind is widely regarded, by Britons especially, as the height of embarrassment. Yet farts are merely the sign of a happily fermenting population of bacteria. Even those of us who never eat a bean emit flatus gas around fourteen times a day. A bit of gas is surely a small price to pay for a healthy garden in our gut.

TOM WAKEFORD, *Liaisons of Life*

Don't faint, but you probably house about two pounds of bacteria in your colon—an impressive intestinal garden of at least 500 species. Your colon bacteria ferment (chemically break down) food residue that your own system can't digest on its own, especially carbohydrates. This yields nutrients for themselves and for you. These microbes also synthesize B vitamins for you; help your body absorb calcium, magnesium, and iron; and fight against competing and invading pathogenic microbes. Although your colon bacteria do great things for you, while breaking down food they produce nitrogen, methane, hydrogen, and a little carbon dioxide that escape your body as . . . flatulence. Sulfate-reducing bacteria produce hydrogen sulfide gas—the "rotten egg" smell that makes breaking wind particularly embarrassing.

Whereas humans rely on bacteria to help extract the nutrients from certain foods they eat, some herbivores owe their very existence to food-digesting microbes. In many respects, grass and leaves are inferior foods. They have low nutrient value and take a long time to digest. Many leaves also contain noxious or toxic chemicals. Cellulose, the carbohydrate that makes up a major part of plants' thick cell walls, contains valuable nutrients. Because vertebrates can't digest cellulose, herbivores rely on microbes such as bacteria, fungi, and protozoans to break down cellulose into digestible nutrients. In contrast, humans—who didn't evolve as strict herbivores—excrete most of the cellulose from the vegetable matter we eat.

Herbivorous vertebrates have evolved two main types of storage vats to house their microbes: hindgut and foregut chambers.

EXAMPLES OF HINDGUT FERMENTERS, those that house symbiotic microbes in their lower intestinal tracts, include horses, elephants, koalas, rabbits, many rodents, several birds, and some lizards, turtles, and fishes. Hindgut fermenters chew their food thoroughly, which breaks the cell walls and releases nutrients.

The cellulose remains undigested until it reaches the caecum (a blind pouch at the junction between the small and large intestine; plural = caeca) and colon, both of which house microbes that ferment cellulose and form volatile fatty acids that the body can absorb.

Koalas eat mainly eucalyptus leaves, high in fiber, low in protein, and packed with toxins. How can they survive on this diet? First, koalas chew their food well—one leaf at a time. Koalas grind a leaf between the molars on one side. Then they flick the wads to the other side and grind some more before swallowing. Second, koalas have long caeca, up to six feet in length. Coarse bits of leaves head down the colon, but finely shredded particles stay in the caecum for about eight days, where bacteria break down the cellulose and extract nutrients, used both by themselves and their host.

Even with these huge fermentation vats, koalas absorb only about 25 percent of the plant fiber they ingest. They store little or no fat. Instead, they conserve energy by sleeping up to 22 hours per day, often propped in the fork of a eucalyptus tree. Koalas move slowly and have a sluggish metabolic rate, so they don't use much energy when they're awake. As with human liver cells that filter out poisons (including industrial chemicals, drugs, food additives, and insecticides) from our blood, a koala's liver neutralizes the plant toxins that enter its bloodstream.

Baby koalas get their intestinal bacteria from their mothers. For about the first seven months, babies only drink their mothers' milk. Then they both nurse and eat "pap," a green fecal fluid their mothers produce that provides bacterial protein and inoculates the baby's caecum with microbes needed for digesting leaves.

Rabbits also have bacteria-loaded caeca. If you've ever watched a rabbit eat, you know it slices through the plant material with its incisors, then grinds the food between its molars. The food spends a few hours in the

stomach, where it undergoes little chemical change; stomach contractions gradually send the food into the small intestine in bursts.

Undigested large particles pass through the colon to be defecated as hard pellets. The smaller particles are shunted into the caecum, where bacteria ferment the food for up to twelve hours. The rabbit's body absorbs some of the fermentation products—amino acids, volatile fatty acids, and vitamins. And, of course, the microbes claim their fair share.

The remaining fermentation products form into soft, moist pellets called "caecotrophs" that pass into the colon and are defecated, typically at night. As caecotrophs emerge from the anus, the rabbit twists around, sucks in the pellets, and swallows without chewing. Caecotrophs are loaded with bacteria, which provide high-value proteins and water-soluble vitamins for the rabbit as the pellets are recycled in its digestive system— an additional source of nutrients. Nursing bunnies acquire their bacteria from eating their mothers' caecotrophs.

Green iguanas, another hindgut fermenter, rely on microbial fermentation in the upper section of the colon to break down the cellulose of fibrous leaves. Their digestive systems are so efficient that a healthy iguana will increase its body mass a hundred-fold from hatching to mature adult—from about 0.4 ounces to about 2.6 pounds—within three years.

During their first week of life, green iguana hatchlings eat soil from the nest chamber and from around the nest. Bacteria in the soil provide the babies with enough help that they can digest plant material during their second week. Needing more bacteria, the hatchlings then climb into the forest canopy. After getting inoculated by eating adults' feces for a few weeks, the young descend to forest-edge vegetation, where they live for several years. Once mature, they reenter the canopy.

CATTLE, SHEEP, DEER, and other grazing ruminants are foregut fermenters with four-chambered stomachs: three that store and process food, and one that digests. The first two chambers are loaded with protozoans, bacteria, and other microbes that ferment cellulose. And do these microbes ever produce the gas! A well-fed lactating cow belches about 130 gallons of methane and other gases per day. In fact, scientists have suggested that cattle and other ruminants may produce more "greenhouse gases" and contribute more to global warming than do all the internal combustion engines of the world!

After Bessie the cow chews a bit on a mouthful of grass, she swallows and the grass passes through the esophagus into the first stomach chamber, the rumen. Rumen microbes ferment the grass, which then passes into the reticulum and softens into a "cud." Several hours later, Bessie's reticulum muscles send the cud back up into her mouth for a second chew.

While ruminating, she mixes the wad with saliva and bacteria and eventually swallows the mush. This time the food passes through the first two chambers into the third, the microbe-filled omasum, for further processing. Finally, the food enters the abomasum, the "true stomach," where the mass mixes with digestive juices. After passing through the small and large intestines, unusable material exits as Bessie's cow pies. Calves get their symbiotic microbes directly from their mothers, but not from eating pies. Instead, cows transfer microbes to their calves when they lick their babies' snouts, and when the calves eat their mothers' regurgitated rumen contents.

As with koalas, rabbits, iguanas, and the microbes that inhabit their guts, both cattle and their symbionts benefit. Microbes enable cattle to extract nutrients from cellulose and survive on grass. The cattle also get vitamins and proteins when they digest some of their symbionts. Microbes rely on their hosts to graze the grass, and the microbes gain nutrients from cellulose fermentation. Some microbes also eat the waste products of other microbes.

Hoatzins, pheasant-sized birds from South America, are one of the few birds that eat mainly leaves. The others—such as ptarmigan, grouse, and ostriches—house their microbes in hindgut fermentation chambers. Hoatzins do it in the foregut. They're "flying cows"! The muscular crop (the bag-like swelling in a bird's esophagus where food is stored temporarily) and lower esophagus together function as a rumen—chambers where microbes ferment plant fiber under constant temperature and acidity conditions. Because these microbes effectively detoxify plant alkaloids, hoatzins can eat leaves of many plants that are poisonous to other herbivores. Fermenting leaves make these birds smell like fresh cow manure, thus their common name of "stinky cowbird." And how do hoatzins get their microbe helpers? For the first three months, the parents regurgitate food from their crops and feed the sticky, greenish goop to their chicks—predigested mash, loaded with microorganisms.

HINDGUT AND FOREGUT fermentation chambers are simply alternative ways of housing the symbiotic microbes that break down cellulose and allow the herbivore host to absorb plant nutrients. Each system has advantages. Hindgut fermenters process food faster than do foregut fermenters. Food passes through a horse in 30 to 45 hours, versus 70 to 100 hours in a cow. Because food passes rapidly, hindgut fermenters can survive on high-fiber, low-quality food—as long as there's plenty of it. This ability explains why wild horses can survive in some areas of the western United States where cattle cannot. On the other hand, foregut fermentation can be highly efficient because microbes work on the plant material *before* it reaches the small intestine, the site of major absorption. Another advantage is that microbes detoxify plant chemicals *before* toxins are absorbed into the bloodstream, not afterward.

Back to us, one last thought. Nutritional experts tell us to EAT LOTS OF FIBER. Why? Our beneficial intestinal microbes thrive on fiber. Dietary fiber such as that found in peas, beans, lentils, and whole grains stays intact in our stomachs and small intestines. Once the fiber passes into the colon, symbiotic bacteria break it down into nutrients that our bodies—and they—can use. The more fiber, the "happier" the bacteria. These bacteria not only help us to assimilate nutrients, synthesize vitamins, and help us absorb minerals; they also produce certain fatty acids, which suppress growth of cancerous cells and reduce the risk of colon cancer. In contrast, certain less helpful bowel bacteria seem to thrive on chemicals found in processed foods. Some of our intestinal bacteria convert sulfate preservatives into sulfides. Not only do we (and others) have to endure the "rotten egg" smell of our farts, but also excessive sulfides can damage cells lining the large intestine, leading to inflammation and ulcers.

The take-home message: To keep beneficial bacteria thriving to the detriment of "less good" bacteria, eat lots of beans and whole grains and cut down on processed food. And never mind that you'll break wind. Microbiologist Tom Wakeford suggests: "Once we understand our gut's microbial friends, perhaps we will learn to love them a little more and blush a little less."

DEADLY DRAGON DROOL

Then Medea led Jason to the sacred grove where the dragon watched beside the Fleece. The huge monster rose up, roaring terribly, as Jason approached. He breathed clouds of smoke and fire and lashed his tail against the oak tree.

MARGARET EVANS PRICE, "Jason and the Golden Fleece," from *Myths and Enchantment Tales*

In folklore and mythology, dragons are often dinosaur-like scaly reptiles that breathe fire, are bad-tempered, and have voracious appetites. In addition to fire, they use long muscular tails, wicked claws, and lethal teeth to defend treasure hoarded at their dens. Wannabe heroes attempt to slay these powerful creatures to regain a kingdom, marry a princess, or secure the wealth.

OUR REAL-WORLD "DRAGON," the Komodo dragon, is the world's largest lizard at 10 feet long and weighing over 200 pounds. A member of the monitor lizard family Varanidae, this gray-brown lizard lives only on Komodo Island and several other Indonesian islands northwest of Australia. The local islanders call their monster *ora* ("grandfather" or "grandmother") or *buaja darat* ("land crocodile"), among other names.

Historians speculate that Chinese traders may have returned home with tales of Komodo dragons as early as the second century A.D., and that these descriptions may have inspired dragon mythology of the Far East. Komodo dragons also may be responsible for the occasional warning ancient cartographers inscribed over the Lesser Sunda Islands of southeastern Asia: "Here there be dragons."

Westerners didn't document Komodo dragons until 1910, when a Dutch first lieutenant named Van Steyn van Hensbroek stationed on the north coast of Flores Island heard about monstrous lizards on the neighboring island of Komodo and set out to verify the tale. He shot a 7-foot lizard on Komodo and sent the skin and a photograph to Peter Ouwens, director of a zoological museum in Java. In 1912 Ouwens named the lizard *Varanus komodoensis*. Shortly afterward, the sultan of one of the islands and Dutch authorities outlawed sport hunting and limited the number of lizards that could be collected for scientific study or zoos. Douglas Burden and others from a 1926 American Museum of Natural History expedition took two live animals and twelve dead specimens back to the New York Zoological Park. In a 1927 *National Geographic* article, Burden was the first to refer to these monitors as "dragon lizards." The name "dragon" stuck.

AS KOMODO DRAGONS galumph along forest trails, they dart their yellow, foot-long, deeply forked tongues in and out of their mouths to smell their environment. These formidable predators have powerful jaw muscles, long claws, and sharp serrated teeth, but they don't fly or breathe fire and they rarely pursue prey. Instead they ambush active animals or attack sleeping prey, including deer, macaque monkeys, and wild boar. A full-grown Komodo dragon can kill a 1,300-pound water buffalo. These lizards also eat carrion.

A Komodo dragon lunges for its prey's abdomen and bites, but often the wounded victim escapes. Although the lizard can run fast, it only runs in short bursts. No problem. The prey will soon drop dead. A Komodo dragon can detect carrion odor from nearly seven miles away. The attacker and likely many other dragons follow the scent and find the animal several days after the bite—dead of septicemia (blood poisoning).

My infatuation with these lizards began in 1968 with a job announcement. The late Walter Auffenberg, herpetologist from the University of Florida, needed a field assistant for his yearlong study of Komodo dragons. I applied—and quickly received my first professional rejection letter. In retrospect, I understand why he wasn't interested in a 115-pound, 22-year-old college coed as a field assistant to study 200-pound dragons. Auffenberg hired a fellow from Bali, who agreed to the job partly because no one else in his graduate class was brave enough to volunteer.

One of the questions Auffenberg addressed during 1969–70 was why the monitors' prey, often large mammals, die so quickly from a single bite. He suspected that because these lizards feast on rotting carcasses, their saliva might contain virulent bacteria. If a Komodo dragon's razor-sharp, one-inch teeth injected pathogenic bacteria, a victim might soon die of infection. Auffenberg collected saliva samples from two monitors, and pathologists from the University of Florida Medical School later identified four bacteria species, all of which could produce severe infections.

That discovery led biologists to wonder why these lizards didn't infect themselves with their own nasty bacteria. Spongy gums cover at least two-thirds of each dragon tooth. When a lizard bites, the gums recede, exposing more of the teeth. Continued rubbing of the gums over the sharp tooth edges causes the gums to bleed, exposing a lizard's bloodstream to the same pathogenic bacteria that kill its prey. Furthermore, interactions among a dozen or more dragons that converge on a carcass aren't always amicable. During fights, the lizards bite each other over access to the feast. One monitor's saliva enters another's bloodstream. Decoding

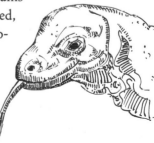

the secret of the lizards' immunity provided the draw for Terry Fredeking, founder of a Texas biotechnology firm, to study the monitors. He wanted to learn if Komodo dragons have natural substances that protect them from their own deadly bacteria. If so, could these be isolated and made into powerful new drugs to fight antibiotic-resistant pathogens in humans?

Collecting dragon drool and blood samples from the dragons' bleeding gums proved to be an adventure in 1995, beginning with an ambush. Fredeking, a team of scientists, and Komodo National Park rangers hid, and when they spotted a Komodo dragon, they slipped a crocodile noose over the lizard's head and pulled tight. Their first subject was an 8-foot 150-pound dragon. Six men jumped onto the angry reptile to immobilize it. Others wrapped duct tape around the head and claws and held down the powerful tail. As Fredeking gloated at having collected saliva with long Q-Tips, one of the crew gasped, "Oh my God!" As described in Michael Shnayerson and Mark J. Plotkin's *The Killers Within*:

Fredeking looked up, and felt the paralyzing fear of the hunter who has gone from being predator to prey. More than a dozen Komodo dragons were advancing from all sides. Drawn by the noisy struggle of the dragon that had been captured, the lizards had converged with the quaintly Komodian hope of eating it—along with the men around it. Panting with adrenaline, the men pushed at the dragons with their forked sticks. With their length, body mass, and sheer reptilian power, the dragons easily could have pushed right up to the men and started chomping away, either at the duct-taped dragon or at the hors d'oeuvre plate of tasty human legs. But the sight of tall men with sticks seems to confuse them. One of the park guards—an old hand at dealing with the dragons— aggressively advanced on one of the larger lizards and pushed him away with his forked stick. For a tense minute or so, the outcome remained uncertain. Then, one by one, the dragons turned and clumped away, including the first creature from which the sample had been taken. Fredeking took a long breath. "Man, oh man," he said. "What we do for science."

Although the crew survived that life-threatening encounter, three men got scratched from sitting on a Komodo dragon's back to restrain it. The scaly skin proved to be rife with pathogenic bacteria, and within hours the men ran fevers. Ciprofloxacin killed their infections—not surprising, since the bacteria had probably never encountered commercial antibiotics.

Fredeking flew the hard-earned saliva and blood samples back to the States. Samples from captive Komodo dragons housed in U.S. and Indonesian zoos provided additional material. Researchers counted 57 species of aerobic (oxygen-requiring) bacteria in the saliva, 54 of which are potentially pathogenic. Analysis of the blood revealed three chemical substances that may hold promise as new antibiotics. Researchers will experiment with mice, then guinea pigs, then primates in a time-consuming process to determine if these antibacterial substances will be useful for humans.

EIGHT YEARS AFTER Auffenberg rejected me as his field assistant, I joined the zoology faculty at the University of Florida. Now my friend and colleague, Walt entertained me with his dragon adventures. He said that although most individuals were shy and fled when disturbed, some acted aggressively without provocation. One particularly aggressive individual entered a tent, stuck its head into a backpack, pulled out a shirt, and tore it to shreds. As Walt told me about getting treed by a large monitor, he laughed and said he was glad the dragon wasn't smaller, given that young monitors climb trees. In fact, for the first few years of life, young Komodo dragons spend much of their time in trees, where they eat insects, geckos, and other small animals. The ground is a danger zone until the young are three feet long because adult Komodo dragons are cannibals.

After hearing about these monitors for years, when I finally had an opportunity to watch a live one at the Shedd Aquarium in Chicago, I was unprepared for the lizard's enormous size, muscular build, and powerful presence. I watched as the dragon lumbered off after a dead rabbit offered by the keeper. As the dragon shook the listless rabbit, I thought how sad that its virulent saliva was now useless to this magnificent beast.

How do Indonesian islanders deal with their dragon neighbors? One anthropologist learned that some villagers believe the monitors are the islanders' siblings. If a lizard is injured, humans will become sick. That belief may offer the dragons some protection. On the other hand, people occasionally kill Komodo dragons that enter their village, arguing that they must protect their children, dogs, and livestock. Komodo dragons occasionally kill people, the most recent being an eight-year-old boy in June 2007. The first dragon-caused human fatality on Komodo Island in 33 years, the boy was going to the bathroom behind a bush when the lizard attacked.

I recently asked Kurt Auffenberg, a teenager when he helped his father in the field during 1969–70, how the locals felt about the monitors. "The villagers just accepted the dragons as animals that lived there with them," Kurt told me. "They didn't necessarily fear them. It was probably like living in India with tigers or in Kenya with lions. There was something

out there that could get you if you weren't careful. But they knew there were many things that could get you—disease, a falling tree, a storm while out fishing, or old age (40 to 50 years). They just worked around the monitors. The villagers buried their dead under hand-hewn lumber so the dragons wouldn't dig up the bodies. You didn't take a nap out in the woods. You usually didn't go out by yourself. You paid attention."

With only 4,000 to 5,000 Komodo dragons left in the wild, the International Union for Conservation of Nature lists the lizards as vulnerable. Komodo National Park, founded in 1980 to protect the monitors, is home to most remaining dragons. Indonesians typically consider Komodo dragons a national treasure. Still, even though it is against the law to kill Komodo dragons, illegal poaching occurs, as well as illegal poaching of the monitors' main prey—deer. Another threat to the dragons' future is loss of habitat due to increased human population growth.

HUMANS HAVE LONG embraced dragons into their literary fantasy world. As we learn more about the pathogenic bacteria living in Komodo  dragon saliva, perhaps science fiction writers will create scaly, bad-tempered creatures that not only breathe fire, but also drool deadly bacteria.

Today's dragon-slaying knights—adventurous biologists—pounce upon 8-foot, 150-pound Komodo dragons. Instead of armor, they carry duct tape. Instead of swords, they wield long Q-Tips. And instead of gathering gold and jewels, they collect blood and saliva, hoping to develop an even more valuable treasure—new drugs to fight antibiotic-resistant pathogens.

MIGHTY MUSHROOMS AND OTHER GOOD FUNGUS AMONG US

Mighty Mushrooms

INGREDIENTS
¾ ounce dried porcini mushrooms
½ cup sliced button mushrooms
½ cup sliced baby portobello mushrooms

½ cup sliced cremini mushrooms
4 ounces oyster mushrooms
4 ounces sliced shiitake mushrooms
⅓ cup olive oil
2½ garlic cloves, crushed
2½ teaspoons ground coriander
kosher or sea salt; freshly ground black pepper
4 tablespoons chopped fresh parsley

1. Soak the dried porcini mushrooms in a little hot water, just to cover, for 20 minutes.
2. In a saucepan, heat the oil and add all the mushrooms. Stir well, cover, and cook gently for 5 minutes.
3. Stir in the garlic, coriander, and salt and pepper to taste. Cook for another 5 minutes until the mushrooms are tender and much of the liquor has been reduced.
4. Mix in the parsley.
5. Serve over noodles, rice, or some other pasta. Enjoy.

After "Mighty Mushrooms," CHRISTINE INGRAM, *Vegetarian and Vegetable Cooking*

As a professor at the University of Florida, I had a poster with a photo of a mushroom taped to my file cabinet. It read: "They keep me in the dark and feed me bullshit." Though many an undergraduate chuckled at the poster, my academic colleagues and I knew firsthand that the message was not entirely facetious.

We associate mushrooms with dark, dank woods, and many do live in such places. Others pop up from cow pies baking in the sun. Sun-loving or shade-loving, mushrooms are simply the spore-containing fruiting bodies that rise up from under the ground—the reproductive structures of mycelia (see introduction to this section). Because mushrooms taste good and are rich in nutrients, they attract hungry animals. The earliest hunter-gatherer people probably collected mushrooms for food. And we have been doing it ever since. Other mushroom-eaters range from handsome and hairy fungus beetles to flying squirrels, tassel-eared squirrels, and kangaroo rats. The advantage to the fungi is that their spores are dispersed when the animals defecate.

For humans, mushrooms are a nearly ideal food: high in protein, B vitamins, minerals, and fiber. They contain all the amino acids essential to human health and are low in calories, virtually free of cholesterol, and easily digestible. Depending on where you live, your supermarket might

offer shiitake, oyster mushrooms, enoki, straw mushrooms, Chinese black mushrooms, portobello, cremini,
porcini, morels, or chanterelles in addition to cultivated
button mushrooms. Mmmm . . . think stroganoff, mushroom risotto, spinach and wild mushroom soufflé,
mushrooms stuffed with seafood . . .

Truffles are fungal fruiting bodies that grow two to fifteen
inches underground on tree roots, especially oaks, willows,
poplars, pines, firs, hickories, and beeches. Red, white, brown,
or black and ranging in size from a kidney bean to a small
potato, truffles of over 100 species grow wild throughout
the world. Their taste varies from delicate to smoky or pungent.
Some smell like fresh earth, wine, garlic, or roasted hazelnuts. Truffles
radiate mystique, in part because we can't predict where or when they will
grow. Besides, how could something that resembles a dried prune dug
from dirt taste so exquisite?

Because they're difficult to find, and because they taste so good, truffles
command a high price. Imagine eating French black truffles, also called
"black diamonds," at $1,000 per pound. Better yet, Italian white truffles
at $2,000 per pound. Or more! In 2007 a three-pound white truffle from
Tuscany in central Italy reportedly fetched $330,000 at auction! Stanley
Ho, an East Asian gambling king, placed the winning bid. The truffle was
served at a 200-guest banquet several days later. Sadly, Ho was not feeling
well and missed the feast.

"Sensual as a kiss" and "dizzying as helium" are descriptors given to
truffles, considered by some to be aphrodisiacs. One believer was Napoleon Bonaparte, self-crowned emperor of France. By 1809 Napoleon had
amassed an immense empire. He worried about the empire's future after
his death, as he had no heirs. He divorced his 46-year-old wife, Josephine,
and in 1810 married 18-year-old Marie Louise, daughter of the emperor
of Austria. Desperately wanting a son, according to legend, Napoleon
turned to truffles. Purportedly, he went on a diet of truffle-stuffed turkey
and champagne. The following year, Marie Louise gave birth to Napoleon
François-Joseph Charles. Truffles may have facilitated the conception, but
they didn't ensure good heath. Napoleon II was a weak child and died of
tuberculosis at age 21.

Truffles' strong odors attract squirrels, chipmunks, and voles. These
and other truffle addicts dig them up, eat them, and disperse spores
through their scats. As we humans have a lousy sense of smell, we use dogs
and pigs to hunt truffles for us. Sows make especially good scouts because

the truffle odor resembles a sex hormone found
in the boar's saliva. Unfortunately, sows love to
eat the truffles they unearth. For this reason,
most *trufficulteurs* prefer to use specially trained
dogs. The dogs ignore the truffles they
find in favor of doggy treats.

Other "good" fungi include little
guys. We harness single-celled yeasts to
make alcoholic drinks. Yeasts feed on sugar from fruit, honey, or grains
and begin the fermentation process: conversion of sugar to ethyl alcohol
and carbon dioxide gas. Depending on the sugar source, the end product
is wine, mead, beer, sake, or some other alcoholic beverage. Although it
wasn't until 1860 that the French chemist Louis Pasteur confirmed that
live yeast organisms convert sugar to ethyl alcohol, over 5,000 years ago
people living along the Nile River knew of the practical side: they brewed
beer. Virtually every human society makes fermented beverages.

Many indigenous peoples of the New World tropical rain forests brew
beer from starchy crops such as corn, peach palm, banana, and *yuca* (a
shrub native to South America that has starchy roots; also called "man-
ioc"). While doing fieldwork in the Ecuadorian Amazon Basin, on ceremo-
nial occasions I drank *chicha* (the local word for brewed "beer") made from
yuca. Quechua women make the fermented brew by chewing hunks of *yuca*
root, then spitting the mass and accumulated saliva into a hollowed-out
gourd. The contents are left for several days to ferment, after which the
fibers are strained out. Enzymes from the saliva turn part of the starch
into sugar—food for the yeasts that cause fermentation. I eventually got
used to the strong and bitter brew, though I always longed for a cold draft
lager in its place.

Elsewhere in the Amazon Basin, *chicha* made from corn is a common
fermented drink. Anthropologist Philippe Erikson lived for a year with the
Matis, a small group of indigenous people from western Brazil. He described the brewing of corn beer by the Matis as follows: After grating the
kernels, women make a cornmeal paste by simmering the ground corn
in water. Once cool, the women chew the paste and then spit it into a
receptacle. As with *chicha de yuca*, enzymes from the saliva convert starch
to sugar, food for the participating yeasts. Every female in the village,
from age four or five and up is expected to help chew and spit—except for
menstruating women. After filtering the mixture, water is added to get
the perfect consistency. A little ripe banana might be added for sweetener,
then the liquid is set aside to ferment.

Yeast is also an essential ingredient in baking leavened bread. The carbon dioxide gas it produces makes the dough rise, and the alcohol evaporates during the baking process. People have been baking leavened bread for thousands of years, relying on the flatulence of a fungus to make dough rise to yield light, fluffy bread.

For at least the past 2,000 years, we've used molds to flavor cheeses. We inoculate *Penicillium roquefortii* into cheese to make blue cheese and Roquefort. As the fungus spreads throughout the cheese, it produces a strong, rich flavor and leaves blue streaks. We don't know who "invented" Roquefort, but folklore claims that a young shepherd boy accidentally left his lunch in a cave near the village of Roquefort, France. When he returned several weeks later, he ate the cheese even though blue veins coursed through it. Amazed at the taste, the boy ran down to the village shouting, "A miracle, a miracle!" The villagers who tried the cheese were equally impressed and soon began storing their own cheese in the caves, which are still used today.

So, why is Roquefort so much more expensive than blue cheese if they're produced by the same fungus? Blue cheese is made with cows' milk, and cows produce a lot of milk. Roquefort is made with sheep's milk. A sheep gives only one quart of milk per day, and then only for six months in a year.

To make Camembert and Brie, we add *Penicillium camembertii* to cheese. This mold grows on the cheese surface, digesting from the outside toward the center, instead of along "cracks" as in Roquefort, ripening the cheese and creating a creamy texture. Curiously, when mold attacks forgotten cheese left in the refrigerator, we amputate the "spoiled" part. But when we ourselves purposely inoculate cheese with mold, we call it "mold-ripened" and adore the stench and unique flavor.

(We also use bacteria to make stinky cheese. Even if you've never smelled Limburger cheese, you've no doubt heard it compared to "stinky feet." Ever wonder why? The same bacterium we use to make Limburger cheese, *Brevibacterium linens*, lives on our skin and is partially responsible for body odor and the peculiar stench of our feet!)

The fungus *Aspergillus oryzae* converts soybeans into a condiment many of us enjoy with Asian food. Fungus is inoculated into pressed cakes of cooked soybeans and wheat flour. After three days the cakes, now covered with yellow fungal growth, are mixed with salt and water, inoculated with a yeast and a bacterium, and left to ferment for six to nine months. Liquid squeezed from the mush is pasteurized, bottled, and sold as soy sauce.

Common fungal parasites called smuts leave masses of black spores on cultivated crop plants, particularly cereal grains. Corn smut causes kernels to swell to many times their normal size and turn silvery-gray. Dark warty masses erupt from the ears—not a pretty sight. Researchers have developed smut-resistant corn hybrids, but some U.S. farmers intentionally grow smutty corn. To find out why, we head south of the border to Mexico. Since pre-Colombian times, *huitlacoche* (pronounced weet-la-KO-chay)— corn smut—has been a delicacy used in Mexican cooking for its earthy flavoring. Mexican farmers plant corn varieties that are especially susceptible to the fungus because smutty ears fetch a higher price than uninfected ears. The buyer gets not only corn but also *huitlacoche.* Mexican food lovers in the United States also enjoy this earthy flavor, which leads some stateside farmers to grow non-smut-resistant corn. The smut that once destroyed the appearance of corn on the cob has redeemed itself. Whereas once it was a farmer's nightmare, it is now a chef's enticement. You can buy *huitlacoche* canned, frozen, or—if you're lucky—fresh.

WE AREN'T THE only beings that benefit from fungi. Let's turn to plants and mycorrhizae. The word "mycorrhiza" comes from the Greek *mykes,* meaning "fungus," and *rhiza,* meaning "root." Mycorrhizae are intimate associations between specialized soil fungi and plant roots. The hyphae of mycorrhizae either wrap around and coat the root surfaces, or they pierce the root and enter its cells.

Of 3,617 species of land plants surveyed, the roots of 80 percent associate with mycorrhizal fungi. Some scientists estimate that at least 90 percent of all land plant species are mycorrhizal. The partnership almost always benefits both parties. Most plant roots can't get enough mineral nutrients on their own, and fungi can't make their own food. Mycorrhizal fungi send out a massive network of hyphae that can extend hundreds of feet beyond their associated tree roots. These hair-like structures constantly "search" for nutrients. More efficient than the plant's own root hairs, the fungal hyphae absorb nutrients, which they share with the plant. Hyphae also supply the plant with water. In return, the plant photosynthesizes and shares carbohydrates with the fungi. These symbiotic soil fungi are so critical to plant nutrition that without them many trees, shrubs, and other land plants would starve. Different trees even share mycorrhizae through common networks. Trees in sunny spots subsidize trees located in deep shade—they share nutrients via the mycorrhizae pipeline.

Mycorrhizae discovered on plant fossils from 400 million years ago are almost identical to present-day mycorrhizae. Scientists believe that the earliest land plants initially got a foothold thanks to "fungus roots." And plant roots and mycorrhizal fungi have been mutually beneficial partners ever since.

Fruiting bodies of mycorrhizal fungi are more visible than their networks of underground hyphae. While walking in damp forest or pasture, you may have seen rings of white or brown mushrooms circling patches of dead grass bordered by lush green grass. English folklore maintained that fairies formed these "fairy rings" by holding hands and dancing in a circle on moonless nights. The mushrooms served as seats for tired fairies. Scandinavians told of elves dancing in circles, leaving their magic behind as mushrooms. In German folklore, the rings were places where witches danced during pagan rituals. Austrian beliefs held that either dragons' breath burned the dead zones, or fiery tails of sleeping dragons scorched the areas.

Though we all love magic, none of the folklore explains the origin of fairy rings. The real explanation, known since the 1790s, is mycorrhizal fungi. Underground hyphae grow out in all directions from the original spore, forming an expanding circle. The hyphae dump digestive enzymes into their environment. These enzymes break down organic matter and release so many nutrients that the plants grow more luxuriantly and turn darker green at the leading edge where the nutrient supply is greatest. Sometimes hyphae become so dense that water cannot penetrate, or the hyphae deplete the soil of nutrients, causing plant life in the center to wither. As the mycelium grows outward, its inner part eventually dies. Periodically the fungus surfaces on the outer edge of the zone as a ring of mushrooms that disperse spores for the massive underground mycelium.

Most fungus-root relationships are mutually beneficial. In a few cases, though, the plant exploits the fungus and gives back nothing in return. One freeloader is Indian pipe, or "ghost plant," found in dense forests in Asia, North America, and northern South America. It resembles the shape of a pipe, has a pure white waxy stem, white scale-like leaves, and a waxy, white bell-shaped flower. Because this plant lacks chlorophyll and cannot photosynthesize, it depends on mycorrhizae to provide it with all the minerals and sugars it needs. You might be thinking, isn't this "the blind leading the blind"? How can mycorrhizal fungi, which lack chlorophyll and normally gain their nutrients from green plants, provide food to plants that also lack chlorophyll? Recall that forest plants are interconnected by networks of mycorrhizae. Indian pipe sponges off mycorrhizae associated with trees and other plants that have chlorophyll. Sugars go from photosynthetic plants through the fungus and into the Indian pipe, so ghost plants ultimately get their energy from photosynthetic forest neighbors. Mycorrhizae are the "middlemen."

WE SOMETIMES OVERLOOK the positive attributes of fungi. But many fungal species sustain life. Some provide basic ingredients; others add spice. Without yeasts, among the smallest of all fungi, life would be dull indeed: no bread, wine, or beer. Mushrooms add zest to our salads, soups, and entrees. Truffles add romantic aroma and flavor to our food, from baked potatoes and scrambled eggs to roasted squab and broiled lobster. Without mycorrhizae, many plants could not develop properly, reproduce, or live at all. The world as we know it would cease to exist. So let's be grateful for the good fungus among us.

BOMBARDED BY BACTERIA

Adam Had 'Em

Amoebas abound in your kisses
And flagellates lurk on your lips.
Your bowels are all swarming with microbes,
B. coli, Giardia and sich . . .
You're nought but a mass of corruption
Passed down from a simian tree
To Adam and Eve and their offspring.
Who says we are equal and free?

ROBERT W. HEGNER, from *Nature Smiles in Verse*

Not all bacteria-human interactions are "good" from the human point of view. A mass of bacterial corruption is how I felt after returning from Latin America, where I had lived for ten days with a runny-nosed, oozy-eyed toddler. Bacteria had taken over my body.

We always harbor bacteria on our skin, in our digestive systems, and in our respiratory tracts. Some live harmlessly in or on our bodies and cause infection only if our resistance is low. Some live on us and we never know it unless they overgrow or produce toxins.

Scientists and physicians generally recognized that bacteria cause disease only about 120 years ago. How different the world has become with that knowledge! Besides regularly washing our hands, now we use sterile procedures in hospitals, eradicate fleas and other agents that carry pathogenic bacteria, fastidiously dispose of sewage, purify our drinking water, pasteurize our milk, quarantine people who have contagious infections, and immunize the healthy.

Fortunately, only a tiny fraction of all bacteria cause human diseases. But they cause some whoppers: leprosy, botulism, tetanus, typhoid fever, cholera, tuberculosis, gonorrhea, syphilis, bubonic plague, Lyme disease, pneumonia, scarlet fever, and meningitis. Pathogenic bacteria are life-forms just "doing their thing" to survive, but in surviving they can kill us.

Pathogenic bacteria affect tissues where they enter the body—skin wounds, the respiratory and gastrointestinal tracts, and the urogenital canal. Many bacteria release enzymes that damage their host's cell membranes. This enzyme-releasing activity kills the surrounding affected cells, and nutrients are released—"food" for the bacteria. Breaking apart the host's cells may also allow the bacteria to spread further through the tissues. If they break open cells of the host's immune system, the bacteria may avoid the host's defenses.

When pathogenic bacteria invade your body, your white blood cells surround and attack them. Antibodies in your blood also attack and kill the foreign invaders. These first-line defenses usually keep bacterial pathogens from establishing themselves. If the pathogens beat off these defenses and begin to spread, though, the affected site might become inflamed. Increased plasma outflow from small blood vessels into the affected tissues causes swelling. The resulting inflammation might be painful, but the benefit to you is that the bacteria are exposed to more antibodies.

Sometimes, though, bacteria win the fight against our bodies' defenses. My ordeal started with catching the toddler's cold just before I left her home. Colds are caused by a virus, so I can't blame that ailment on bacteria. The cold inflamed the mucous membranes in my sinuses. My nasal passages swelled. The mucus couldn't drain properly, creating an ideal condition for normally harmless bacteria to multiply. Eventually the bacteria invaded my sinuses, and soon after returning home I developed a terrific sinus infection with all the classic symptoms: pounding headache, sore jaws, toothache, and sore throat from postnasal drip. I dragged myself to the walk-in medical clinic. An hour later I swallowed my first dose of Amoxicillin, confident the antibiotic would kill the invaders. Wanting to know who to blame, I Googled "sinusitis bacteria" and learned that the bacteria most commonly implicated in sinus infections are *Streptococcus pneumoniae, Haemophilus influenzae, Moraxella catarrhalis,* and *Staphylococcus aureus.* Which species, I wondered, had infected me? Or all of them?

The following morning, I couldn't open my eyes—my eyelids were stuck together. With the help of a wet washcloth I washed the yellow gunk loose and peered into the mirror. No doubt about it. I had conjunctivitis— pinkeye. Back to the clinic, where the same doctor prescribed antibiotic eye drops.

Knowing I was highly contagious, I hunkered down at home for the next week, feeling like Typhoid Mary, the first known carrier of typhoid fever in the United States. Although Mary Mallon from Ireland had recovered from typhoid fever, appar- 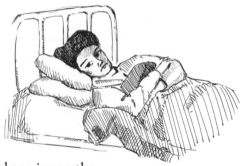 ently her gallbladder was still infected with typhoid bacteria. Mary had lots of contact with the public, as she was a cook in New York City. When Mary refused to have her gallbladder re- moved, she was imprisoned. Later she agreed to quit her job as cook and was released. But she changed her name and continued cooking—and passing on the bacteria. Authorities hospitalized her to remove her from the general population. And there she stayed until her death more than 20 years later. Mary infected at least 53 people between 1900 and 1915, three of whom died from the disease. No, I didn't want to be "Conjunctivitis Marty."

From Google I learned that the most common species to cause pinkeye are *Streptococcus pneumoniae, Haemophilus influenzae,* and *Staphylococcus aureus*—three of the four that commonly cause sinus infections. Eight days into my ten-day course of Amoxicillin, my sinus infection felt no better. The bacteria were partying, apparently resistant to the antibiotic. Back to the clinic. After joking that my presence was becoming a habit, the doctor prescribed a different antibiotic—Levaquin. Within a few days I could breathe through my nose again. My headache vanished. All my nasty bacteria were beaten—so I thought.

Three days before leaving Latin America, I had traipsed through a weedy field, searching for puddles that might house tadpoles. I found tadpoles, but chiggers in the grass found me. The tiny, red mite larvae inserted their mouthparts, injected saliva, and sucked my fluids. The next morning my legs burned with itchy red welts. I slathered on anti-itch gel and tried not to scratch. Five weeks later, back in Arizona, the welts still itched, and they began to erupt. The lesions oozed and became honey-yellow and crusty. New itchy spots sprang up on my arms and legs—small bumps full of fluid, surrounded by circles of reddened skin. The doctor took one look at my legs and pronounced it impetigo.

Impetigo! The word brought back unpleasant memories, and I shared with the doctor my childhood experience with impetigo. When I was twelve years old, I developed an oozy, crusty sore on my arm. Leprosy! I was sure of it. Terrified of being sent to a leper colony, I kept my plight secret until my mother noticed the sore. Within hours she had me in the doctor's office, and I was soon applying an antibiotic ointment against impetigo. My current doctor laughed. "I don't think it's leprosy this time either." He prescribed more antibiotics—this time Cefuroxime.

Another Google: "impetigo bacteria." I learned that impetigo is caused by *Staphylococcus aureus* and/or *Streptococcus pyogenes*—the same ol' *Staphylococcus* plus a new entrant onto my bacterial scene. These bacteria can live harmlessly on your skin, but if you get a skin break—such as an insect bite—they invade. Mine was a classic case. If left untreated, my impetigo bacteria could have affected my kidneys, bones, joints, and lungs, and even have caused blood poisoning. The next day I checked out several books on microbiology from the university library, itching to learn more about *Streptococcus* and *Staphylococcus* bacteria. What I learned nearly transformed me into a germ-control freak.

Humans serve as the reservoir for *Staphylococcus aureus*. People carry these staph bacteria asymptomatically in their noses, skin, ears, throats, armpits, hair, and gastrointestinal and urogenital tracts. In some places the bacteria occur in up to 70 percent of people. Bacterial carriers may be symptom-free, but they can infect themselves and others. A person might snort and blow the bacteria out of the nose, or sneeze and spray droplets. You can pick up staph bacteria from a towel, doorknob, cell phone, handshake, a kiss. Anything you touch or breathe in. People also carry *Streptococci* (strep) bacteria without showing symptoms of disease. These bacteria can persist for weeks after leaving the human body. They settle on surfaces and in dust waiting for unsuspecting victims.

MY RECENT EXPERIENCE with "bad" bacteria has made me increasingly grateful for antibiotics—drugs that kill or stop the growth of bacteria by interfering with the cells' normal processes. Antibiotics are natural products. Bacteria and fungi release antibiotics into their environment to inhibit other microbes that compete with them for space and other resources—the microbial version of chemical warfare. In effect, then, we exploit some microbes' defenses to fight against other microbes that infect us. Scientists have identified over 8,000 naturally produced antibiotics, but we currently use only about 60 to treat human diseases. After growing cultures of the antibiotic-producing microbes, lab technicians transfer the material to huge fermentation vats. The microbes grow quickly in these conditions. Eventually, the antibiotic substance is extracted from the culture and purified.

But as we all know, antibiotic-resistant bacteria are an increasing problem. When you take an antibiotic, the defenseless bacteria die. Bacteria that can resist the antibiotic live and multiply. Bacteria have various ways of fighting back. Some produce enzymes that dismantle the antibiotics. Others change their cell walls so the antibiotics can't bind and destroy the walls. Still others develop slightly altered ribosomes, the tiny bodies within cells that make proteins. This alteration is enough that the antibiotics can no longer prevent the bacteria from producing the proteins vital to their existence. Whatever the mechanism, antibiotic-resistant survivors multiply—potentially a millionfold in 24 hours—and are then passed on to other people. The more bacteria are exposed to antibiotics, the faster they evolve resistance. By 2008, 20,000 deaths due to drug-resistant staph bacteria were reported in the United States for the previous year.

Currently, one particularly grave problem is methicillin-resistant *Staphylococcus aureus,* or MRSA. This germ mainly causes skin infections, but it can be life-threatening if it gets into your bloodstream or lungs. Pneumonia, sinus infections, and "flesh-eating" wounds caused by MRSA are becoming more common. These drug-resistant bacteria are causing outbreaks of staph infections in schools, hospitals, jails and prisons, gymnasiums and locker rooms, and in other social situations where people are in close contact with each other. An estimated 53 million people worldwide are thought to carry MRSA in their noses or on their skin. The ones carrying it asymptomatically are healthy, but they infect others. In 2005 the

Centers for Disease Control and Prevention estimated that the number of MRSA infections treated in U.S. hospitals was 278,000. The current number of annual deaths in the United States caused by MRSA exceeds the number of deaths caused by AIDS.

We can't get away from pathogenic bacteria. We can't outrun them, and we can't outsmart them. The trick is to coexist by keeping our immune systems strong with a healthy diet and plenty of sleep. As our mothers taught us, we need to wash our hands often and keep our fingers out of our mouths. And it doesn't hurt to minimize contact with sick toddlers—if that's realistic . . .

A CLOAK OF ANTIBIOTICS

The trouble with being a hypochondriac these days is that antibiotics have cured all the good diseases.

CASKIE STINNETT

In the last essay, we saw how humans use antibiotics—defenses of bacteria and fungi—to destroy our own disease-causing microbes. Other organisms do the same—and without the help of giant pharmaceutical firms. Certain insects use bacterial antibiotics to destroy pathogenic fungi. The following stories of antibiotic-coated insects also illustrate how our understanding of the natural world changes as we continue to observe and experiment. The thrill of discovery is a driving force for biologists, whether in the field or in the laboratory.

WHEN MY FAMILY and I lived in the cloud forest community of Monteverde, Costa Rica, we planted zinnias for our four-year-old daughter.

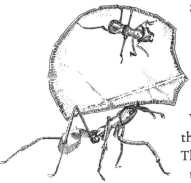 Soon after the tender leaves unfurled, leaf-cutting ants invaded. Leaf-cutters scissored off pieces of leaves with their mandibles and carried them back to their underground nest. These leaves became the substrate for growing a spongy, bread-like fungus. Worker ants feast on the terminal bulbs of the hyphae that grow on the leaf fragments. They also feed the bulbs to their queen and to their larvae.

Workers tend their fungus garden with as much diligence as we did our garden—though in very different ways. The ants chew the leaf particles, then add saliva and feces to form a sticky mass. Next they pluck tufts of fungal threads from another part of the garden and "plant" the fungus in the newly prepared mush. Periodically the ants add more fecal fertilizer.

At the time the ants destroyed our zinnias in 1986, biologists had long known that both leaf-cutters and fungi benefit from their relationship. The ants gain food, and the fungi gain a nutritious substrate and tender loving care. The fungi also gain a free ride to new locations, because when a young queen ant leaves her colony to start a new nest, she carries a wad of fungus in her mouth pouch. With this she plants a new garden.

The story gets better. Fungi release enzymes onto their food. The enzymes break down the proteins and carbohydrates into simple compounds that the fungal hyphae can absorb. With the leaf-cutting ants' particular brand of fungi, the protein-digesting enzymes are locked in the terminal bulbs—which the ants eat. But the fungi need their enzymes, so how do they get them back? The enzymes recycle through the ants' feces. Fortunately for the fungi, the enzymes are not broken down in the ants' digestive tract. Thus, the fertilizer the ants spread on the fungi is essential, enabling the fungi to digest nutrients from the leaves with their own enzymes. This recycling is also critical for the ants, because if the fungi couldn't produce the terminal bulbs, the ants would starve.

And better still. As farmers, leaf-cutting ants battle a different fungus—a parasitic one—that invades their gardens. This "enemy" fungus is so deadly that it can potentially wipe out a garden in a few days, but the ants control it with still another organism's help. Part of a leaf-cutting ant's cuticle (outer layer of its body) bears a powdery, whitish-gray crust. Depending on the ant species, this crust sits just below the mouth on what would be the neck if an ant had a neck, or under the front legs. Biologists hadn't paid much attention to this crust, assuming it was just excretion from the cuticle. In 1999 Cameron Currie and three colleagues reported the crust to be a mass of filamentous bacteria in the genus *Streptomyces*—soil-living microbes that produce antibiotics.

Tests revealed that the ants' *Streptomyces* produce antibiotics that strongly suppress growth and spore germination of the "enemy" fungus. Of 22 species of leaf-cutting ants surveyed, every one wore a cloak of *Streptomyces*. How do the bacteria get into a new colony? After examining females and males during mating flights, the research team found that none of the males had the bacteria, whereas all queens carried the bacteria on their cuticles.

Instead of the "simple" story of the ants and the cultivated fungus living in a two-way mutualistic relationship, we now know that four organisms interact: ants, cultivated fungi, parasitic fungi, and antibiotic-producing bacteria. The ants and cultivated fungi share a mutualistic relationship. The *Streptomyces* bacteria benefit both the ants and their cultivated fungi when they suppress the parasitic fungi. The bacteria themselves gain by getting dispersed to new locations, and they presumably gain nourishment from the ants. Three "good" guys against one "bad." Currie and his colleagues suggested that this four-part interaction has evolved over the past 50 million years—for nearly as long as leaf-cutting ants have been gardening.

(As an aside, leaf-cutting ants aren't the only invertebrates that farm fungus. Some termites and beetles also cultivate fungi for their own nutrition. In 2003 another fungus farmer was added to the list: a marine snail that lives in salt marshes along the Atlantic coast of North America. There, salt marsh periwinkles actively graze live salt marsh cordgrass. Instead of eating the leaves, however, the periwinkles make wounds on the leaf surfaces, preparing substrate for intertidal marsh fungi to grow. Even more amazing, these snails fertilize the fungus-invaded wounds with their nutrient-rich fecal pellets. Days later the snails eat the fungi. Who knows . . . perhaps future fieldwork will reveal that these salt marsh periwinkles are also cloaked in antibiotic-producing bacteria that kill off a yet-to-be discovered parasitic fungus that attacks their leaf fungal gardens. Two decades ago we never would have suspected such a relationship in leaf-cutting ants!)

THE INSECT PROTAGONIST in the second story is a solitary yellow-and-black hunting wasp called the European beewolf. A female beewolf, ½- to ¾-inch long, digs a nesting burrow in sandy soil. Then she

hunts honeybees, which she stings, paralyzes, and stores in her burrow. Once she has stored several paralyzed honeybees, she excavates a side burrow ending in a separate chamber called a "brood cell." She hauls her incapacitated victims to the brood cell, lays an egg on one of the bees, and closes the side burrow. She continues gathering and paralyzing honeybees and digging side burrows and brood cells until she has laid all her eggs and provisioned them with food. The larvae hatch after two or three days. They eat the paralyzed honeybees for another five to eight days and then spin cocoons, from which wasps emerge nine months later.

Most brood cells lie eight to fourteen inches beneath the ground surface, where conditions are warm with nearly 100 percent humidity—ideal conditions for fungal growth. Curiously, though, fungal infestation is uncommon. Why? Some scientists assumed that venom from the wasp's sting protects the bees from fungi. In 2001 Erhard Strohm and Eduard Linsenmair compared the time for fungus to attack the following: (1) bees the investigators killed by freezing, (2) bees paralyzed by wasps but not provisioned in brood cells, and (3) bees paralyzed and provisioned by wasps. Fungus attacked every freeze-killed bee within three days. About a day later, fungus began attacking the paralyzed-only bees and by the fifth day had infested all of these. In contrast, 50 of the 54 paralyzed and provisioned bees remained fungus-free. Even the other four were not attacked until much later than the other groups. How were the wasps protecting the bees from fungus through their provisioning activity? The investigators observed the wasps licking the entire body surface of the honeybees and concluded that female beewolves apply an anti-fungal chemical to their prey.

Martin Kaltenpoth and three colleagues, including Erhard Strohm, soon added another piece to the puzzle. They wondered how cocooned wasp larvae survive nine months in the brood cell without getting attacked by fungi. The researchers again focused on the mother beewolf because before laying an egg, she smears a large amount of white secretion from her antennae onto the ceiling of the brood cells. Molecular analysis of DNA from the females' antennae revealed—what else?—a new species of *Streptomyces* bacteria.

Could this be another insect with antibiotic protection? Indeed, experimental results strongly suggested that the high concentration of bacteria on the cocoon walls help protect the larvae from fungal attack. Whereas 15 of 18 larvae exposed to the substance emerged or survived at least until the end of the experiment at 45 days, only 1 out of 15 larvae without access to the white substance emerged from its cocoon. How do the female wasps get the bacteria in the first place? They likely pick up the bacteria directly from their mothers. Larval wasps eat some of the white substance before spinning their cocoons.

What's in it for the *Streptomyces*? Like the bacteria associated with leaf-cutting ants, the wasps' bacteria gain dispersal and presumably nutrition.

Curious whether this beewolf species was unique in housing *Streptomyces* bacteria, the investigators examined two other wasp species from the same genus—one from southern Europe and the other from North America. Both species carried *Streptomyces* bacteria in their antennae, suggesting that associations between protective bacteria and ground-nesting wasps—and perhaps other ground-dwelling arthropods—may be widespread.

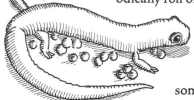

LIKE GROUND-NESTING INSECTS, other animals that live underground in warm, humid conditions are vulnerable to pathogenic fungi. Frogs and salamanders that brood their eggs in underground nests periodically roll or turn their eggs. The consensus among herpetologists used to be that this manipulation destroyed fungal hyphae and/or prevented developmental abnormalities or yolk layering. Now we know that at least some underground-brooding amphibians have antibiotic-producing bacteria living on their skin. By jostling the eggs, the parents presumably transfer bacteria from their skin to their eggs. Recently, scientists have discovered antifungal bacteria on pond-breeding amphibians as well and have speculated that antibiotic-producing bacteria may protect many amphibians—not just those that brood their eggs underground in fungi-infested nests.

Now, how about those marine snails? Could they also be cloaked with fungi-fighting bacteria? We're just beginning to explore this fascinating world of interactions between big and little organisms. The thrill of discovery beckons.

INVASION OF THE BODY SNATCHERS

There is an ancient legend told in the Himalayas, relating the way *Cordyceps* [a parasitic fungus] was originally found; it was from a time long ago, when tribes people of Tibet and Nepal took their animals into the high mountain pastures for springtime grazing. There they would see goats and yaks grazing on some sort of a small, brown grass-like mushroom, growing from the head of a caterpillar. After eating this strange looking creature, the animals would become frisky and start chasing the other goats and yaks around with lustful intent. I guess this added vigor must have looked like a pretty good thing to those tribes' people, so they started collecting these small mushrooms and eating them as well. They got frisky as well, and even a bit lustful, or so the story goes. . . .

JOHN HOLLIDAY AND MATT CLEAVER, "On the Trail of the Yak"

Enough about "good" and "bad" interactions. Now for the downright "scary"! As I describe in *In Search of the Golden Frog:*

My hand brushes against my stomach and I feel peach fuzz. More fuzz on my thigh. What's happening to my body? I grab my flashlight and peer under the covers. GROSS! My entire body is covered with greenish gray fungus. I try to scrape it off with my Swiss Army knife. The long filaments fall off, but there's a stubble left, stubbornly adhering to my skin. I peer into my pocket mirror and scream in horror at the fuzzy green monstrous face peering back at me.

My scream woke me from the nightmare. I was a graduate student and had been living in the hot, humid jungle of eastern Ecuador for almost nine months. Fungus covered my leather shoes and belt, permeated my clothes and sheets, infested my American dollars and Ecuadorian sucre bills, and invaded a patch of skin on my arm. Even worse had been the small horrors I'd witnessed during the preceding month.

While searching for frogs in the rain forest, I had come face-to-face with dead beetles, flies, wasps, ants, moths, and other insects posed rigidly on leaves or the ends of twigs in tortured, contorted positions. Cottony white or gray filaments smothered most victims' bodies, and a dusting of white filaments radiating outward made it seem as if someone had pinned the insects down with tiny nets. Stiff black threads ending in pinhead knobs sprouted from other corpses. The bodies resembled victims in a science fiction movie. Attacked by parasitic fungi, the "possessed" insects had experienced what must have been, from my perspective, a terrifying death.

The gruesome Ecuadorian horror show begins when a spore from a parasitic fungus attaches to an insect's cuticle, the outer layer of its exoskeleton. The spore germinates and a "germ tube" penetrates into the host's body, growing into thousands of thread-like hyphae. The fungus infiltrates the host's tissues and feeds on nutrients. Toward the end, the insect scrambles about erratically, eventually climbs onto a leaf, twig, or other exposed surface, then dies. The exoskeleton breaks open, and hyphae smother the rigid cadaver. Fungal rhizoids (root-like anchoring hyphae) attach the victim's body to the leaf, and an elongated fruiting body sprouts from the corpse. The knob at the top of the fruiting body produces spores, which disperse in the wind or are carried away by other animals.

An insect infected with fungi often behaves in a particular way that increases spore dissemination. In essence, the fungus manipulates the host's behavior to its advantage. When investigators studied ants in Switzerland, they concluded that fungal hyphae found in the brain and nervous system of infected ants disrupted the ants' normal behavior. The infected ants climbed up grasses and latched on tightly with their legs and mandibles before dying. In that elevated position, spores more easily dispersed into the environment. Ground-dwelling ponerine ants from Africa rarely climb until they're infected with parasitic fungus, then they begin climbing like seasoned rock-climbers. When convulsions cause the ants to lose their grip and fall to the ground, they gamely retrace their steps upward. Eventually they seize the vegetation with their legs and mandibles and die. By modifying their hosts' behavior, parasitic fungi both increase their own reproductive success and ensure their long-term survival. The cadaver attached to a leaf or stem means that the fungus can produce fruiting bodies over extended periods of time—depending on the fungal species, from hours or days to three months. If the host died on the ground, it might get buried by leaf litter or stomped on before the fruiting bodies could sprout.

My field biologist husband, Pete, recently returned from Cuba with a similar tale. Near dusk he and his Cuban colleagues were in the Zapata Swamp in the south-central part of the island. Small bromeliads of all sizes festooned the trees from waist level on up. Someone asked, "What are these little black dots on the leaf tips?" Then, "Hey, they're tiny dead ants!" Almost every bromeliad leaf tip had at least one ant victim. One tiny bromeliad supported over 50 corpses. The ants appeared to be hunched over in agony and were clearly dead. Pete suggested that if his friends looked closely they might see a tiny stiff thread with a bulb-like structure at the end, emerging from each ant. Sure enough, under a microscope each ant had what Pete called the "Fungus from Hell"—miniature version—emerging from its head or trunk.

The next morning one Cuban described his nightmare of being invaded by a fungus that forced him to climb a tall tree, froze him into position, and then erupted from his head. From then on, the fungus was a great source of jokes. Whenever a person said something stupid, the others teased, "Hey, is that a fungus sticking out of your head?" And when my husband stumbled while lecturing in Spanish, he'd exclaim, "Uh-oh, it's started," freeze himself into an awkward, grotesque position, and hold a finger above his head.

AND THEN THERE are the body snatchers from parts of Asia. *Dong chong xia cau*, the Chinese name for the caterpillar fungus (*Cordyceps sinensis*), literally means "winter worm, summer grass" because caterpillars seen in winter appear to turn into plants during the summer. This fungus para-

sitizes insect larvae, mainly caterpillars of a ghost
moth on the Tibetan Plateau of Tibet, Nepal, and
part of China. These caterpillars live beneath the
surface of the ground, where they bore into tree
roots. The caterpillar fungus enters its host, germinates, and mummifies the body by feeding on
nutrients and replacing the host's tissues with its
hyphae. After killing the caterpillar, the fungus
sends out a dark brown or black fruiting body that ends
in a club-like cap. The fruiting body winds its way upward toward
the light, surfaces aboveground, and releases spores.

Although the behavior of these fungi is gruesome from our point of
view, species in the genus *Cordyceps* contain valuable pharmacological
properties and have long been used for traditional medicines. As described in the quote, according to folklore Himalayan herders discovered
the medicinal benefits of the caterpillar fungus 1,500 to 2,000 years ago
by watching their goats and yaks become energized after grazing on the
fungal fruiting bodies sprouting from caterpillar cadavers. The fungus
has been used in traditional Chinese medicine to treat a variety of ailments, including hiccups, loss of appetite, asthma, fatigue, hemorrhoids,
impotence, cancer, and kidney, lung, and heart disease. Many people also
use the fungus as a daily energizing tonic. The infected caterpillars, often
called "vegetable caterpillars," are thought to have an excellent balance of
yin and yang—the two life forces that flow through the human body—
because they seem to be both plant and animal.

When the snow melts in the spring, local people search the Tibetan
Plateau and collect the mummified caterpillar carcasses and the fungus
"sprouts." Some people brew tea from the fruiting bodies; others cook
them with meat, traditionally duck, or add them to soup. The fungus is
currently popular among Asians as an herbal remedy to boost energy,
strengthen the body after exhaustion or extended illness, and prolong life.

The caterpillar fungus has become popular outside of Asia ever since
September 1993, when two Chinese athletes, Wang Junxia and Qu Yunxia,
broke several world records at the Chinese National Games in Beijing.
Wang ran the 10,000-meter race 42 seconds faster than any previous
woman and broke the women's world record in the 3,000-meter run as
well. Qu set a new women's world record in the 1,500-meter race. The
athletes had trained hard, and on the advice of their coach they drank a
stress-relieving tonic made from the caterpillar fungus. Chinese women
endure intense training schedules, running the equivalent of a marathon

(26.2 miles) each day for six months of the year. Western runners who have tried training at this intensity have broken down from stress. How do the Chinese athletes do it? Advocates claim it's the caterpillar fungus tonic.

Many pharmacies in the Chinatowns of large U.S. cities sell blocks of mummified caterpillars and attached fungus, but the fungus is also cultivated on soybeans or in a liquid nutrient broth, to satisfy the increasing demand. Health food stores and online distributors sell tablets, capsules, and liquid drops—all less expensive and potentially more aesthetically pleasing to consume than the fungus au naturel. Just look for the word "Cordyceps." Whereas once the fungus was an exclusive herbal reserved for emperors and nobles, now you can buy a bottle of 90 capsules for $20. The product is thought to be safe, with only a small percentage of people experiencing side effects of dry mouth, diarrhea, or nausea. The most commonly reported side effect is increased libido— which elicits few complaints! *Cordyceps sinensis* probably works by increasing blood flow and oxygen supply to the liver and other organs, which helps the body use energy efficiently. To be safe, consult your doctor before adding this to your pharmacopoeia.

For those addicted to dark chocolate, the company Fungi Perfecti (www.fungiperfecti.com) sells CordyChi chocolate—a "delicious indulgence" that combines dark chocolate, freeze-dried *Cordyceps sinensis*, and reishi mushroom mycelium. Their chocolate drops or bars are available in seven flavors, including chocolate blueberry crunch, chocolate mint crisp, chocolate mocha, and chocolate coconut almond. Or you can pamper yourself with a mug of steaming organic hot chocolate laced with *Cordyceps* and reishi.

On the other hand, if you prefer to order a block of mummified caterpillar carcasses, will you smell death when you open your package? No. Depending on your particular sensory organs, the caterpillars and their growths might smell faintly sweet, smoky, spicy, or bittersweet. Why don't the dead caterpillars smell like death? *Cordyceps* fungi often live in moist soil normally conducive to microbial decay. The fungi produce antimicrobial compounds that keep their hosts from rotting. Scientists have isolated the antibacterial agent cordycepin from one species of *Cordyceps*, and an antifungal agent from another. Thanks to these agents, soil microorganisms such as bacteria and other kinds of fungi can't plunder the nutrient reserves the *Cordyceps* needs to produce its stalked fruiting bodies and spores.

Research is currently under way to determine if *Cordyceps* might provide us with new antibiotics and other medicines. The journey from traditional Oriental remedies to Western-style drugs is long and complex, in part due to the cultures' contrasting philosophies of medicine and science. Oriental scientists and doctors tend to seek a holistic understanding of illness, considering all the many factors that lead to imbalance of the body. Western scientists and physicians often try to understand each aspect of illness separately. In order for natural products to be turned into medicines, Western scientists isolate, identify, and purify the target agents. Although we know that certain fungal compounds contain medicinal properties—and that for at least 1,500 years people have used fungi to improve their quality of health and endurance—Western scientists are just beginning to convert these compounds into drugs.

ONE LAST TIP. Fungi Perfecti also sells full-spectrum mushroom formulas for dogs. Available in a powder to be mixed with wet food (MUSH) or a biscuit (Muttrooms), the products contain five certified organic mushrooms: *Cordyceps sinensis*, reishi, maitake, shiitake, and turkey tail. Powder or biscuit, the combination of fungi is formulated to "help promote superior health for your dog." I plan to buy some for Conan's next birthday. Yin and yang. Plant and animal. Some fungal body snatchers are both killers and healers.

BODY SNATCHERS REVISITED

It was the common belief in Athens that whoever had been taught the Mysteries would, when he died, be deemed worthy of divine glory. Hence all were eager for initiation.

SCHOLIAST ON ARISTOPHANES, in *The Frogs*

When I was twelve, we played the game of "witches' brew" at my Halloween party. As the ten of us kids huddled under a blanket, Mom passed us ingredients for our brew (actually more of a stew): cats' eyeballs (peeled grapes), rabbit turds (raisins), skunk intestines (cooked spaghetti), and human brains (globs of Jell-O). Passing and handling each one in the dark, we groaned and shrieked, "eeuw, gross!" Mom popped each ingredient into a bowl once it had made the circuit. The game ended when we each ate a bite of our "stew-brew" under the darkness of the blanket.

Later, Shakespeare's *Macbeth* gave me a new perspective on witches' brews. The first witch placed a toad in the kettle. The second witch added

"eye of newt," "toe of frog," "tongue of dog," and "lizard's leg," among
other goodies. The third witch added "scale of dragon," "tooth of wolf,"
and other ingredients to complete the gruel. Unlike most other ingredients that were animal body parts, tooth of wolf is actually a folk name for
ergot—a pharmacologically active fungus.

What exactly is ergot—or "tooth of wolf," if you prefer Shakespeare?
And what is its place in history other than an ingredient in witches' brew?
Bear with me, and you'll see that witches crop up again—this time with-
out their brew.

LET'S BEGIN WITH the quote. In religion, "mysteries" are secret
ceremonies. Only persons belonging to the group or soon to be initiated
are allowed to witness or participate in the ceremony. One of the most
famous of the ceremonies in ancient Greece was the Eleusinian Mysteries.
Any person who spoke Greek was allowed to join, even a woman or slave,
as long as the person had not committed murder. Among those initiated
were the philosophers Aristotle and Plato and the writers Sophocles and
Cicero. Promised happiness and a special life after death, cult members
celebrated an annual autumn event in the city of Eleusis near Athens to
worship Demeter, goddess of agriculture, fertility, and grain. The Eleusinian Mysteries flourished for nearly 2,000 years, until Christians uprooted
the cult about A.D. 395. Forbidden under penalty of death to reveal the
teachings, members guarded the secrets so well that they have been lost
forever. We do know, however, that one important part of the ceremony
was drinking *kykeon*, a sacred concoction that cleared the members' souls,
prepared them for inevitable death, and brought about intense visions. An
anonymous poet, writing in the seventh century B.C., revealed the ingredients of *kykeon* as water, barley, and a fragrant mint.

The ancients who celebrated the Eleusinian Mysteries may have had
more in mind than merely securing happiness in the afterlife. In the 1970s
the late Swiss chemist Albert Hofmann and others speculated that drinking *kykeon* produced hallucinogenic effects from ergot—a parasitic fungus
that grows on barley. Eleusinian priests may have collected fungus from
barley and other grasses growing near the temple and purposely added it
to the *kykeon*.

SPORES OF ERGOT FUNGI germinate on flower stigmas of wild
and cultivated grasses and then grow into the ovaries. These plant parasites are body snatchers, absorbing nutrients and replacing the grain's
seeds with their purplish-black curved masses of compacted cells called

"sclerotia." The word "ergot" comes from the French *argot*, meaning "a cock's spur," in reference to the appearance of the sclerotia. Like many other fungi that parasitize plants, grain ergots affect our lives on the grand scale by reducing crop yields. On the small scale, when ingested, ergots may profoundly affect the human body and mind.

The ergot fungus *Claviceps purpurea* lives on wheat, barley, and other cereal grains, especially rye. The purplish-black structures (the sclerotia) the fungus leaves behind in place of the developing seeds contain potent chemicals, including alkaloids. These alkaloids protect the sclerotia, the fungal resting stage, from being eaten by insects. The sclerotia overwinter, germinate in the spring, and form tiny mushroom-like structures that produce spores, continuing the life cycle.

Baking does not destroy ergot alkaloids. When people eat bread made from infected grain, they get ergot poisoning, or ergotism, an illness that has plagued people for centuries. Between A.D. 800 and 900, thousands of peasants from the Holy Roman Empire ate bread made from infected rye grain and died. From A.D. 994 to 995, as many as 20,000 people—about half the local population—died from ergot poisoning in the Aquitaine region of what is now southern France. France was a center of ergotism because rye served as a staple food for the poor, and the cool, wet climate favored fungal infection.

In 1670 a French physician speculated that ergot-infected rye caused the poisoning. He tried to convince people not to eat bread made from infected grain, but the poor had few other options and the deaths continued. Two hundred years after the French physician's warning, ergot was demonstrated to have caused the poisoning that took so many lives. Even though we now have better methods of cleaning grain, the twentieth century has seen epidemics—11,300 reported cases in Russia from 1926 to 1927 and 200 cases in England in 1927. In 1951 some 230 cases in Provence, France, resulted in 32 instances of insanity and 4 deaths.

Ergot alkaloids produce two distinct forms of toxic reactions. Some alkaloids cause neurological dysfunction, leading to convulsions. The victim twists and contorts in pain, trembling and shaking, and may sense ants crawling under the skin. Sometimes delusions and hallucinations follow. Other alkaloids cause narrowing of the blood vessels leading to extremities. As blood flow is decreased, infection occurs in the extremities, accompanied by burning pain. This constriction can lead to gangrene, and the victim may lose earlobes, fingers, toes, arms, or legs. The extremity falls off without pain or loss of blood. This type of ergotism was commonly called "holy fire," because victims assumed that their agony was retribution for their sins. It was also called "St. Anthony's fire" because the order of St. Anthony, founded in 1093, cared for ergotism victims during the Middle Ages.

THE SALEM WITCHCRAFT TRIALS, held in 1692 in the farming community of Salem Village in the Massachusetts Bay Colony, comprise a tragic chapter of American history. In late December 1691, eight girls, including the nine-year-old daughter and eleven-year-old niece of the minister, broke out in blasphemous screaming fits during which they uttered strange sounds, experienced convulsive seizures, crawled under the furniture, threw objects, melted into trance-like states, contorted themselves into odd positions, and complained of being pricked with pins and cut with knives. The Massachusetts Bay Colony was made up of settlers who had brought with them from England a strong belief in witchcraft. When doctors found no physical evidence of illness, one physician concluded the afflicted had been bewitched.

To identify the witches, a witch cake was baked using rye meal and the girls' urine. After consuming the cake, the victims claimed that three women had bewitched them. One was the minister's slave from Barbados, another an elderly beggar who often muttered under her breath, and the third an irritable old woman of ill repute. All were jailed in late February. The girls did not improve, and they later accused others to be witches. By the time the trials ended in late fall 1692, 19 people (14 women and 5 men) had been convicted of witchcraft and been hanged. Another man was pressed to death with large stones when he refused to enter a plea to the charge of witchcraft. Another 175 to 200 suspected witches were imprisoned, and 5 died while incarcerated.

In 1976 Linnda Caporael, a behavioral psychologist, published a paper in the journal *Science* suggesting that convulsive ergotism may have caused the behavior exhibited by the "possessed" girls in Salem Village. The symptoms of people who have eaten ergot-contaminated grains were consistent with the behaviors noted in the records of the Salem witchcraft trials—hallucinations, delusions, crawling sensations on the skin, vertigo, headaches, vomiting, and violent muscle spasms. The ergot fungus thrives during rainy springs and summers, such as those of 1691, and rye was the staple grain of the village at the time. The rye grains consumed when the first unusual behaviors appeared could have been infected with ergot. The following summer was dry, and no one else in Salem Village developed the characteristic symptoms. If ergot on rye (sounds like a bad sandwich . . .) were the culprit, why didn't more people in the village show symptoms of ergotism during the winter of 1691–92? Ergotism is a complicated ailment, and its virulence depends on how extensively the grain is infected, how much and for how long the person eats contaminated grain, and individual sensitivity. Caporael concludes her paper by stating: "Without knowledge of ergotism and confronted by convulsions, mental disturbance, and perceptual distortions, the New England Puritans seized upon witchcraft as the best explanation for the phenomena."

THE ERGOT FUNGUS that robs plants of nutrients, snatches the seeds' bodies, and kills us can also heal. Chemists have intensively studied ergot alkaloids and have isolated and modified these compounds for various medicines, including drugs to treat migraines and to stem hemorrhaging.

Early attempts to treat migraine headaches included bloodletting and drilling a hole in the person's skull to release evil spirits responsible for the pain. During the nineteenth century, physicians began using ergot to control migraines. Some patients found relief; others experienced symptoms of ergot poisoning. Because some patients benefited from ergot, chemists attempted to isolate the useful alkaloids. In 1920 ergotamine was isolated, and drugs containing this alkaloid still provide relief for some migraine sufferers. Ergotamine is thought to work by stimulating serotonin (a chemical needed to transmit nerve signals to the brain), constricting the blood vessels around the brain, and decreasing inflammation.

Chinese writings dating back to 1100 B.C. reveal the use of ergot in obstetrics. In 370 B.C. Hippocrates noted the use of ergot to stop postpartum hemorrhage. Because epidemics of ergotism were associated with frequent miscarriages among pregnant women, European midwives reasoned that ergot causes uterine contractions. Cautiously, the midwives administered preparations of ergot to women having prolonged labor with unproductive contractions, a use first documented in 1582. In 1808 an American physician reported on the use of ergot to "quicken childbirth." The preparation became popular, but sometimes resulted in powerful contractions that asphyxiated the fetus and resulted in a stillbirth, ruptured the woman's uterus, and even killed the mother. An investigation followed, and by 1824 the use of ergot to stimulate uterine contractions was recommended only to stem postpartum hemorrhage. Because the crude preparations resulted in undefined dosage and safety margins, by the late nineteenth and early twentieth century chemists worked to isolate specific alkaloids that might prove useful. Two alkaloids became standard drugs for this purpose, but in 1935 a water-soluble extract of ergot, called ergometrine, was isolated and developed into a better drug for stimulating uterine contractions. Although

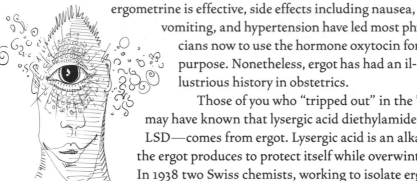

ergometrine is effective, side effects including nausea, vomiting, and hypertension have led most physicians now to use the hormone oxytocin for this purpose. Nonetheless, ergot has had an illustrious history in obstetrics.

Those of you who "tripped out" in the '60s may have known that lysergic acid diethylamide—LSD—comes from ergot. Lysergic acid is an alkaloid the ergot produces to protect itself while overwintering. In 1938 two Swiss chemists, working to isolate ergot alkaloids, synthesized LSD. One, the late Albert Hofmann, discovered its hallucinogenic properties in 1943 when he accidentally ingested some. It was based on this research that Hofmann speculated, along with an ethnobotanist and an eminent Greek scholar, that *kykeon* may have been laced with ergot.

The trip that consuming LSD produces is a mental and emotional experience that lasts from eight to twelve hours. Users undergo a roller-coaster ride of emotions and sensory perception, and they have out-of-body experiences. Sensations vary from exuberance and positive energy (a "good trip") to intense fear and depression (a "bad trip"). Timothy Leary, a lecturer in clinical psychology at Harvard University, advocated LSD as a way to examine one's spirituality by gaining new insights into past experiences. Love, peace, and tripping out defined life in the '60s and '70s for heavy

LSD users. An indirect product of ergot, through the LSD route, has reached many millions of people from diverse cultures if you believe the speculation that the Beatles' song *Lucy in the Sky with Diamonds*, with its "tangerine trees and marmalade skies," refers to a "good" LSD trip.

So, how many of Aristotle's and Plato's epiphanies came while tripping out on sacred *kykeon*? The ancient Greeks knew about the psychoactive properties of ergot, which also makes one wonder if the famous philosophers and writers tripped out more often than once a year during the autumnal celebration of the Eleusinian Mysteries—a mystery we'll probably never answer.

AND WHAT ABOUT the recipients of ergotized witches' brew? Do they happily trip out or writhe in convulsions? Either way, it's a good thing Mom never put "tooth of wolf" in our Halloween stew!

The essence of ergot is the alkaloids packed in the structures that take over the seeds' bodies—chemicals that protect the vulnerable resting stage from hungry insects. When we eat the alkaloids, they can "snatch" our brains, altering our perception of who we are, even why we are. On the upside, they free us of earthly notions, awareness, and thinking. We can go somewhere else, be someone else. On the downside, the alkaloids cause convulsions, lead to gangrene, and even kill. But we have turned them around to our advantage. We've isolated and incorporated individual ergot alkaloids into powerful medicines—even one that precludes the need to drill holes in our skulls to release the evil spirits responsible for migraines.

Conserve Interactions, Not Just Species

IMAGINE A 70-FOOT cottonwood tree growing in moist soil next to a stream in the western United States. This tree, with its spreading, leafy crown, offers more than just picnic shade. Aphids tap sap from leaf veins. Leaf-hoppers pierce stems and suck out the juices. Leaf-chewing beetles and leaf-roller cater-pillars munch leaves. Beetles live and feed under-neath the bark. These insect herbivores harm the tree, but many will get their comeup-pance. Black-capped chickadees forage for insects along the branches, and predaceous lacewing larvae snarf up the aphids—thus helping out the tree.

Imagine now that a pair of eager beavers moves into the watershed where the cottonwood tree grows. In a short time, the beavers will profoundly alter the habitat for those animals living in, on, and underneath that cotton-wood. Beavers are "ecosystem engineers": they directly or indirectly con-trol the availability of space, food, or other resources by physically chang-ing the environment. They will eat all the leaves from the lower branches of the cottonwood, resulting in less food for the insect herbivores. If the insects aren't there, the insectivorous birds will go elsewhere for dinner. If the insectivorous birds aren't around, the hawks will go elsewhere . . . and so on with cascading effects.

The actions of one species often affect many others. No species lives in a vacuum, nor does any pair of species. When one species goes extinct or increases or decreases in density, it most likely affects a cloud of others. At some point the beavers will cut down the tree. Their woodcutting and dam-building activities will cause long-term changes in the landscape, modifying interactions between and among innumerable other species, from bacteria and fungi to mammals.

Still, the activities of one beaver pair don't hold a candle to what a human can do with a chain saw or bulldozer. Thanks in no small part to those human tools, the world is experiencing the early phases of what conservation biologists call the sixth major extinction episode in the Earth's history, one that could easily surpass the mass extinction that erased the dinosaurs 65 to 70 million years ago. Some of the world's most respected conservation biologists warn that by the year 2100 two-thirds of all plant and animal species could be headed for extinction.

As the most profound ecosystem engineers on Earth today, we humans need to understand our impact and accept the responsibility for preserving the diversity of life. Ethics may be the most powerful reason to do so. Each living organism has intrinsic value and a right to exist. In 1982 the World Charter for Nature, adopted by the United Nations General Assembly and signed by over 100 nations, stated: "Every form of life is unique, warranting respect regardless of its worth to man, and, to accord other organisms such recognition, man must be guided by a moral code of action."

After reading about the amazing interactions herein, though, I hope you will agree with what ecologist Daniel Janzen urged over 30 years ago: we need to conserve interactions, not just species. The World Charter for Nature might better have stated: "Every form of life *and every interaction* [are] unique, warranting respect regardless of its worth to man. . . ."

All life is in some way interconnected. If we destroy unique relationships between and among organisms by causing the extinction of individual species, we will lose as we occupy an increasingly desolate world. Just think what wondrous interactions are yet to be discovered. That thought alone should make each of us a conservationist.

Glossary

alkaloid: Class of chemical compound found in certain plants and animals; serve to protect the organism from being eaten.

allogrooming: Mammals grooming (cleaning) other individuals of the same species.

alloparental care: Taking care of young (of the same species) that are not one's own.

allopreening: Birds preening (cleaning) other individuals of the same species.

antibiotics: Products released by bacteria and fungi into their environment to inhibit other microbes that compete with them for space and other resources; we have developed drugs for our own use from certain natural antibiotics.

Beltian bodies: Small food bodies on acacia trees and certain other plants; rich in proteins, lipids, and carbohydrates, they provide food for ants.

bromeliad: Member of a large New World tropical and subtropical family of plants; usually with radiating clusters of long, narrow leaves; rainwater may collect in the centers of the leaf clusters.

brood cell: Separate chamber as in a female solitary wasp's burrow, where one egg is laid.

brood parasitism: Laying eggs in another animal's nest and leaving the parenting to the host.

brood reduction: Elimination of some individuals within a brood or clutch.

commensalism: Symbiotic relationship in which the member of one species benefits and the other is neither helped nor harmed.

cross-pollination: Transfer of pollen from the stamen of one plant to the pistil of another plant.

ectoparasite: Parasite that lives on the outside of its host's body.

elaiosome: Fatty edible appendage found on certain seeds; "reward" for ants that disperse seeds.

epiphyte: Plant that grows on a tree branch or trunk of another, larger plant; epiphytes get water and food from the air and from decaying organic matter near their roots (they are not parasites); also called "air plants."

fitness, inclusive: Sum of an individual's fitness based on the relative number of genes it passes on directly through descendants plus the genes it passes on through relatives other than direct descendants.

fitness, relative: Ability of an individual to contribute its genes to the population; individuals that leave behind the greatest number of descendants have the highest fitness relative to other individuals in the population.

frugivore: Fruit-eating animal.

herbivore: Plant-eating animal.

honeydew: Sugary liquid excreted by certain sap-feeding insects such as aphids.

hyphae (singular = hypha): Threadlike cells of multicellular fungi; these filaments absorb nutrients from their immediate environment.

inquiline: Animal that lives in the occupied dwelling place of an animal of another species.

insectivore: Insect-eating animal.

kleptoparasitism: Stealing food from another animal.

labellum: Central petal of an orchid; has an unusual shape that often includes a landing platform for insect pollinators; also called the lip.

monogamy: Prolonged association between one male and one female; association lasts at least long enough to raise one brood; does not imply sexual fidelity.

mutualism: Symbiotic relationship in which members of both species benefit.

mycelium: Collective mass of hyphae of a multicellular fungus.

mycorrhizae: Intimate associations between soil fungi and plant roots.

myrmecophile: Organism that spends at least part of its life cycle with ants; thrives in association with ants; "ant-loving."

myrmecophyte: Plant that has specialized hollow structures that shelter ants; lives in obligatory relationship with ants.

myrmecotrophy: Ants "feeding" plants.

nectary: Tiny, cup-like nectar-producing gland found on certain plants.

nematocyst: Harpoon-like capsule in the stinging cell of a sea anemone and certain other related animals; injects toxin when it discharges.

ovipositor: Egg-laying organ of many female insects, used to insert eggs into soil, wood, fruits, leaves, and the bodies of other animals; females of certain fishes also have ovipositors.

parasitism: Symbiotic relationship in which the member of one species benefits at the expense of the other species; often does not result in death, at least not quickly.

phoresy: One species of animal riding on another species to get to food, mates, or for some other reason.

pollinarium: Composite structure of pollinia (pollen sacs) and associated appendages; found in orchids.

pollinia: Sacs containing thousands of pollen grains; found in orchids.

pupate: To become a pupa (temporary, inactive stage during which the insect's adult body form develops, as in the chrysalis of a butterfly).

rhizome: Horizontal plant stem that grows at or just below ground level.

symbiosis: Close relationship in which one species lives in, on, or with another species; three main types are commensalism, mutualism, and parasitism.

trophallaxis: Exchange of liquid food between individual ants in a colony.

ungulate: Mammal whose toes end in a hoof; even-toed ungulates include pigs, deer, cattle, sheep, antelope, hippopotamuses, giraffes; odd-toed ungulates include horses, rhinoceroses, and tapirs.

References Consulted and Suggested Reading

Not Tonight, Honey

Arnqvist, G., and L. Rowe. *Sexual Conflict.* Princeton, NJ: Princeton University Press, 2005.

Barry, D. *Dave Barry's Guide to Marriage and/or Sex.* Emmaus, PA: Rodale Books, 2000.

Crump, M. L. "Aggression in Harlequin Frogs: Male-Male Competition and a Possible Conflict of Interest between the Sexes." *Animal Behaviour* 36 (1988): 1064–77.

———. *In Search of the Golden Frog.* Chicago: University of Chicago Press, 2000.

Judson, O. *Dr. Tatiana's Sex Advice to All Creation.* New York: Metropolitan Books, Henry Holt, 2002.

Koprowski, J. L. "Removal of Copulatory Plugs by Female Tree Squirrels." *Journal of Mammalogy* 73 (1992): 572–76.

Morrow, E. H., and G. Arnqvist. "Costly Traumatic Insemination and a Female Counter-Adaptation in Bed Bugs." *Proceedings of the Royal Society of London* B 270 (2003): 2377–81.

Reinhardt, K., R. Naylor, and M. T. Siva-Jothy. "Reducing a Cost of Traumatic Insemination: Female Bedbugs Evolve a Unique Organ." *Proceedings of the Royal Society of London* B 270 (2003): 2371–75.

Stutt, A. D., and M. T. Siva-Jothy. "Traumatic Insemination and Sexual Conflict in the Bed Bug *Cimex lectularius.*" *Proceedings of the National Academy of Sciences* 98 (2001): 5683–87.

To Have and to Hold

Avital, E., and E. Jablonka. *Animal Traditions: Behavioral Inheritance in Evolution.* Cambridge: Cambridge University Press, 2000.

Black, J. M. "Introduction: Pair Bonds and Partnerships." In *Partnerships in Birds: The Study of Monogamy,* ed. J. M. Black, pp. 3–20. Oxford: Oxford University Press, 1996.

Black, J. M., S. Choudhury, and M. Owen. "Do Barnacle Geese Benefit from Lifelong Monogamy?" In *Partnerships in Birds: The Study of Monogamy,* ed. J. M. Black, pp. 91–117. Oxford: Oxford University Press, 1996.

Choudhury, S. "Divorce in Birds: A Review of the Hypotheses." *Animal Behaviour* 50 (1995): 413–29.

Emlen, S. T., and L. W. Oring. "Ecology, Sexual Selection, and the Evolution of Mating Systems." *Science* 197 (1977): 215–23.

Ens, B. J., S. Choudhury, and J. M. Black. "Mate Fidelity and Divorce in Monogamous Birds." In *Partnerships in Birds: The Study of Monogamy*, ed. J. M. Black, pp. 344–401. Oxford: Oxford University Press, 1996.

Fisher, H. E. *Anatomy of Love: The Natural History of Monogamy, Adultery, and Divorce.* New York: W. W. Norton, 1992.

Gowaty, P. A. "Battles of the Sexes and Origins of Monogamy." In *Partnerships in Birds: The Study of Monogamy*, ed. J. M. Black, pp. 21–52. Oxford: Oxford University Press, 1996.

Gowaty, P. A., and D. W. Mock. "Avian Monogamy." *Ornithological Monographs* 37 (1985): 1–121.

Mock, D. W., and M. Fujioka. "Monogamy and Long-Term Pair Bonding in Vertebrates." *Trends in Ecology and Evolution* 5 (1990): 39–43.

Orell, M., S. Rytkonen, and K. Koivula. "Causes of Divorce in the Monogamous Willow Tit, *Parus montanus*, and Consequences for Reproductive Success." *Animal Behaviour* 48 (1994): 1143–54.

Russell, E., and I. Rowley. "Partnerships in Promiscuous Splendid Fairy-Wrens." In *Partnerships in Birds: The Study of Monogamy*, ed. J. M. Black, pp. 162–73. Oxford: Oxford University Press, 1996.

Shuster, S. M., and M. J. Wade. *Mating Systems and Strategies.* Princeton, NJ: Princeton University Press, 2003.

Wittenberger, J. F., and R. L. Tilson. "The Evolution of Monogamy: Hypotheses and Evidence." *Annual Review of Ecology and Systematics* 11 (1980): 197–232.

You Scratch My Back, I'll Scratch Yours

Gaston, A. J. "Social Behaviour within Groups of Jungle Babblers (*Turdoides striatus*)." *Animal Behaviour* 25 (1977): 828–48.

Gumert, M. D. "Payment for Sex in a Macaque Mating Market." *Animal Behaviour* 74 (2007): 1655–67.

Hausfater, G., and R. Sutherland. "Little Things that Tick Off Baboons." *Natural History* 93 (February 1984): 55–61.

Nelson, H., and G. Geher. "Mutual Grooming in Human Dyadic Relationships: An Ethological Perspective." *Current Psychology* 26 (2007): 121–40.

Smuts, B. "What Are Friends For?" *Natural History* 96 (February 1987): 36–45.

Sparks, J. H. "Flock Structure of the Red Avadavat with Particular Reference to Clumping and Allopreening." *Animal Behaviour* 12 (1964): 125–36.

Bubble Blowers, Pothole Plugs, and Other Group Hunting Roles

Allen, D. L. "How Wolves Kill." *Natural History* 88 (May 1979): 46–51.

Anderson, C., and N. R. Franks. "Teams in Animal Societies." *Behavioral Ecology* 12 (2001): 534–40.

Baker, C. S., and L. M. Herman. "Whales that Go to Extremes." *Natural History* 94 (October 1985): 52–61.

Fanshawe, J. H. "Serengeti's Painted Wolves." *Natural History* 98 (March 1989): 56–67.

Gazda, S. K., R. C. Connor, R. K. Edgar, and F. Cox. "A Division of Labour with Role Specialization in Group-Hunting Bottlenose Dolphins (*Tursiops truncatus*) off Cedar Key, Florida." *Proceedings of the Royal Society: Biological Sciences* 272 (2005): 135–40.

Gorman, M. L., M. G. Mills, J. P. Raath, and J. R. Speakman. "High Hunting Costs Make African Wild Dogs Vulnerable to Kleptoparasitism by Hyenas." *Nature* 391 (1998): 479–81.

Greenough, J. W. "Whales at Table." *Natural History* 90 (December 1981): 30–35.

Knopf, F. L. "A Pelican Synchrony." *Natural History* 85 (December 1976): 49–57.

Powell, S., and N. R. Franks. "How a Few Help All: Living Pothole Plugs Speed Prey Delivery in the Army Ant *Eciton burchellii*." *Animal Behaviour* 73 (2007): 1067–76.

Ternes, A. P. "Picnic *à la dauphine*." *Natural History* 95 (April 1986): 70–73.

Uetz, G. W. "Sociable Spiders." *Natural History* 92 (December 1983): 63–68.

Vucetich, J. A., R. O. Peterson, and T. A. Waite. "Raven Scavenging Favours Group Foraging in Wolves." *Animal Behaviour* 67 (2004): 1117–26.

Wilson, E. O. *Sociobiology: The New Synthesis.* Cambridge, MA: Belknap Press of Harvard University Press, 1975.

The Babysitters' Club

Bombeck, E. *Motherhood, the Second Oldest Profession.* New York: McGraw-Hill, 1983.

Hrdy, S. B. "Liquid Assets: A Brief History of Wet-Nursing." *Natural History* 104 (December 1995): 40.

Hurxthal, L. M. "Our Gang, Ostrich Style." *Natural History* 95 (December 1986): 34–40.

Jouventin, P., C. Barbraud, and M. Rubin. "Adoption in the Emperor Penguin, *Aptenodytes forsteri*." *Animal Behaviour* 50 (1995): 1023–29.

Lee, P. C. "Allomothering among African Elephants." *Animal Behaviour* 35 (1987): 278–91.

Rood, J. P. "What Ever Happened to the C Pack Sisters?" *Natural History* 97 (February 1988): 41–47.

Roulin, A. "Why Do Lactating Females Nurse Alien Offspring? A Review of Hypotheses and Empirical Evidence." *Animal Behaviour* 63 (2002): 201–8.

Sibley, D. A. *The Sibley Guide to Bird Life and Behavior.* New York: Alfred A. Knopf, 2001.

Woolfenden, G. E., and J. W. Fitzpatrick. *The Florida Scrub Jay: Demography of a Cooperative-Breeding Bird.* Princeton, NJ: Princeton University Press, 1984.

———. "Sexual Asymmetries in the Life History of the Florida Scrub Jay." In *Ecological Aspects of Social Evolution*, ed. D. I. Rubenstein and R. W. Wrangham, pp. 87–107. Princeton, NJ: Princeton University Press, 1986.

Sound the Alarm!

Blumstein, D. T., and K. B. Armitage. "Alarm Calling in Yellow-Bellied Marmots: I. The Meaning of Situationally Variable Alarm Calls." *Animal Behaviour* 53 (1997): 143–71.

Blumstein, D. T., J. Steinmetz, K. B. Armitage, and J. C. Daniel. "Alarm Calling in Yellow-Bellied Marmots: II. The Importance of Direct Fitness." *Animal Behaviour* 53 (1997): 173–84.

Dominey, W. J. "Mobbing in Colonially Nesting Fishes, Especially the Bluegill, *Lepomis macrochirus*." *Copeia* (1983): 1086–88.

Francis, A. M., J. P. Hailman, and G. E. Woolfenden. "Mobbing by Florida Scrub Jays: Behaviour, Sexual Asymmetry, Role of Helpers and Ontogeny." *Animal Behaviour* 38 (1989): 795–816.

Graw, B., and M. B. Manser. "The Function of Mobbing in Cooperative Meerkats." *Animal Behaviour* 74 (2007): 507–17.

Manser, M. B. "The Acoustic Structure of Suricates' Alarm Calls Varies with Predator Type and the Level or Response Urgency." *Proceedings of the Royal Society of London, Series B* 268 (2001): 2315–24.

McGowan, K. J., and G. E. Woolfenden. "A Sentinel System in the Florida Scrub Jay." *Animal Behaviour* 37 (1989): 1000–1006.

An Intimate Act

DeNault, L. K., and D. A. McFarlane. "Reciprocal Altruism between Male Vampire Bats, *Desmodus rotundus*." *Animal Behaviour* 49 (1995): 855–56.

Hölldobler, B., and E. O. Wilson. *The Ants.* Cambridge, MA: Belknap Press of Harvard University Press, 1990.

———. *Journey to the Ants: A Story of Scientific Exploration.* Cambridge, MA: Belknap Press of Harvard University Press, 1994.

Hoyt, E. *The Earth Dwellers: Adventures in the Land of Ants.* New York: Simon & Schuster, 1996.

Wilkinson, G. S. "Bat Blood Donors: Feeding and Sharing in Vampire Bat Colonies." In *The New Encyclopedia of Mammals*, ed. D. Macdonald and S. Norris, pp. 766–67. Abingdon: Andromedia Oxford Limited, 2001.

———. "Food Sharing in Vampire Bats." *Scientific American* (February 1990): 76–82.

———. "Reciprocal Food Sharing in the Vampire Bat." *Nature* 308 (1984): 181–84.

———. "Social Grooming in the Common Vampire Bat, *Desmodus rotundus*." *Animal Behaviour* 34 (1986): 1980–89.

———. "The Social Organization of the Common Vampire Bat: I. Pattern and Cause of Association." *Behavioral Ecology and Sociobiology* 17 (1985): 111–21.

Whatever Happened to Baby Booby?

Crump, M. L. "Cannibalism in Amphibians." In *Cannibalism: Ecology and Evolution among Diverse Taxa*, ed. M. A. Elgar and B. J. Crespi, pp. 256–76. New York: Oxford University Press, 1992.

———. "Cannibalism by Younger Tadpoles: Another Hazard of Metamorphosis." *Copeia* (1986): 1007–9.

Drummond, H., and C. G. Chavelas. "Food Shortage Influences Sibling Aggression in the Blue-Footed Booby." *Animal Behaviour* 37 (1989): 806–19.

Gerhardt, R. P., D. M. Gerhardt, and M. A. Vasquez. "Siblicide in Swallow-Tailed Kites." *Wilson Bulletin* (1997): 112–20.

Harano, K., and Y. Obara. "The Role of Chemical and Acoustical Stimuli in Selective Queen Cell Destruction by Virgin Queens of the Honeybee *Apis mellifera* (Hymenoptera: Apidae)." *Applied Entomology and Zoology* 39 (2004): 611–16.

Hayssen, V. D. "Mammalian Reproduction: Constraints on the Evolution of Infanticide." In *Infanticide: Comparative and Evolutionary Perspectives*, ed. G. Hausfater and S. B. Hrdy, pp. 105–23. New York: Aldine, 1984.

Mock, D. W., and B. J. Ploger. "Parental Manipulation of Optimal Hatch Asynchrony in Cattle Egrets: An Experimental Study." *Animal Behaviour* 35 (1987): 150–60.

Pfennig, D. W., H. K. Reeve, and P. W. Sherman. "Kin Recognition and Cannibalism in Spadefoot Toad Tadpoles." *Animal Behaviour* 46 (1993): 87–94.

Pfennig, D. W., P. W. Sherman, and J. P. Collins. "Kin Recognition and Cannibalism in Polyphenic Salamanders." *Behavioral Ecology* 5 (1994): 225–32.

Springer, S. "Oviphagous Embryos of the Sand Shark, *Carcharias taurus*." *Copeia* (1948): 153–57.

Stevens, L. "Cannibalism in Beetles." In *Cannibalism: Ecology and Evolution among Diverse Taxa*, ed. M. A. Elgar and B. J. Crespi, pp. 156–75. New York: Oxford University Press, 1992.

Wilson, E. O. *The Insect Societies*. Cambridge, MA: Belknap Press of Harvard University Press, 1971.

Hunting Partners

Bshary, R., A. Hohner, K. Ait-el-Djoudi, and H. Fricke. "Interspecific Communicative and Coordinated Hunting between Groupers and Giant Moray Eels in the Red Sea." *Public Library of Science Biology* (December 5, 2006). http://biology.plosjournals.org/perlserv/?request=get-document&doi=10.1371/journal.pbio.0040431&ct=1.

Diamond, J. "Strange Traveling Companions." *Natural History* 97 (December 1988): 22–27.

Downer, J. *Weird Nature: An Astonishing Exploration of Nature's Strangest Behavior*. Willowdale, ON: Firefly Books, 2002.

Hoh, E. "Flying Fishers of Wucheng." *Natural History* 107 (October 1998): 66–71.

Isack, H. A., and H.-U. Reyer. "Honeyguides and Honey Gatherers: Interspecific Communication in a Symbiotic Relationship." *Science* 243 (March 10, 1989): 1343–46.

Minta, K. A., and S. C. Minta. "Partners in Carnivory." *Natural History* (June 1991): 60–63.

Minta, S. C., K. A. Minta, and D. F. Lott. "Hunting Associations between Badgers (*Taxidea taxus*) and Coyotes (*Canis latrans*)." *Journal of Mammalogy* 73 (1992): 814–20.

Rasa, O. A. "A Taru Life Story." *Natural History* 94 (September 1985): 36–41.

Smith, B. D. "Fish Tales." *Wildlife Conservation* (February 2007): 26–31.

Willis, E. O., and Y. Oniki. "Birds and Army Ants." *Annual Review of Ecology and Systematics* 9 (1978): 243–63.

Taken to the Cleaners

Beebe, W. *Galápagos: World's End*. New York: G. P. Putnam's Sons, 1924.

Breitwisch, R. "Tickling for Ticks." *Natural History* (March 1992): 57–63.

Darwin, C. *The Voyage of the Beagle*. London: Penguin, 1989. Originally published by Henry Colburn, 1839.

Foster, S. A. "Wound Healing: A Possible Role of Cleaning Stations." *Copeia* (1985): 875–80.

Johnson, W. S., and V. C. Chase. "A Record of Cleaning Symbiosis Involving *Gobiosoma* sp. and a Large Caribbean Octopus." *Copeia* (1982): 712–14.

Limbaugh, C. "Cleaning Symbiosis." *Scientific American* 205 (1961): 42–49.

Sazima, I., and R. L. Moura. "Shark (*Carcharhinus perezi*), Cleaned by the Goby (*Elacati-*

nus randalli), at Fernando de Noronha Archipelago, Western South Atlantic." *Copeia* (2000): 297–99.

Street, P. *Animal Partners and Parasites.* New York: Taplinger, 1975.

Vogt, R. C. "Cleaning/Feeding Symbiosis between Grackles (*Quiscalus:* Icteridae) and Map Turtles (*Graptemys:* Emydidae)." *Auk* 96 (1979): 608–9.

She's Got a Ticket to Ride

Ashe, J. S., and R. M. Timm. "Predation by and Activity Patterns of 'Parasitic' Beetles of the Genus *Amblyopinus* (Coleoptera: Staphylinidae)." *Journal of Zoology* (London) 212 (1987): 429–37.

Colwell, R. K. "Stowaways on the Hummingbird Express." *Natural History* 94 (July 1985): 57–63.

O'Toole, B. "Phylogeny of the Species of the Superfamily Echeneoidea (Perciformes: Carangoidei: Echeneidae, Rachycentridae, and Coryphaenidae), with an Interpretation of Echeneid Hitchhiking Behaviour." *Canadian Journal of Zoology* 80 (2002): 596–623.

Silva, J. M., Jr., and I. Sazima. "Whalesuckers on Spinner Dolphins: an Underwater View." JMBA2 *Biodiversity Records* (2006): 1–6. www.mba.ac.uk/jmba/pdf/5201.pdf.

Timm, R. M., and J. S. Ashe. "The Mystery of the Gracious Hosts." *Natural History* 97 (September 1988): 6–10.

Vander Meer, R. K., D. P. Jouvenaz, and D. P. Wojcik. "Chemical Mimicry in a Parasitoid (Hymenoptera: Eucharitidae) of Fire Ants (Hymenoptera: Formicidae)." *Journal of Chemical Ecology* 15 (1989): 2247–61.

Zeh, D. W., and J. A. Zeh. "Novel Use of Silk by the Harlequin Beetle-Riding Pseudoscorpion, *Cordylochernes scorpioides* (Pseudoscorpionida, Chernetidae)." *Journal of Arachnology* 19 (1991): 153–54.

———. "On the Function of Harlequin Beetle-Riding in the Pseudoscorpion, *Cordylochernes scorpioides* (Pseudoscorpionida: Chernetidae)." *Journal of Arachnology* 20 (1992): 47–51.

Houseguests, Unlike Dead Fish, Don't Always Smell in Three Days

Anker, A., G.-V. Murina, C. Lira, J. A. Vera Caripe, A. R. Palmer, and M.-S. Jeng. "Macrofauna Associated with Echiuran Burrows: A Review with New Observations of the Innkeeper Worm, *Ochetostoma erythrogrammon* Leuchart and Rüppel, in Venezuela." *Zoological Studies* 44 (2005): 157–90.

Berrill, N. J., and J. Berrill. *1001 Questions Answered about the Seashore.* New York: Dodd, Mead, 1957.

Carr, A. *A Naturalist in Florida: A Celebration of Eden.* New Haven, CT: Yale University Press, 1994.

Carson, R. *The Edge of the Sea.* Boston: Houghton Mifflin, 1955.

Gray, I. E. "Changes in Abundance of the Commensal Crabs of *Chaetopterus*." *Biological Bulletin* 120 (1961): 353–59.

Hunt, R. H. "Toad Sanctuary in a Tarantula Burrow." *Natural History* 89 (March 1980): 49–53.

Lips, K. R. "Vertebrates Associated with Tortoise (*Gopherus polyphemus*) Burrows in Four Habitats in South-Central Florida." *Journal of Herpetology* 25 (1991): 477–81.

Preston, J. L. "Communication Systems and Social Interactions in a Goby-Shrimp Symbiosis." *Animal Behaviour* 26 (1978): 791–802.

Yanagisawa, Y. "Strange Seabed Fellows." *Natural History* (August 1990): 46–51.

Be It Ever So Humble

Borror, D. J., and R. E. White. *A Field Guide to the Insects of America North of Mexico.* Boston: Houghton Mifflin, 1970.

Crump, M. L., and J. A. Pounds. "Lethal Parasitism of an Aposematic Anuran (*Atelopus varius*) by *Notochaeta bufonivora* (Diptera: Sarcophagidae)." *Journal of Parasitology* 71 (1985): 588–91.

Cruz, Y. P. "The Defender Role of the Precocious Larvae of *Copidosomopsis tanytmemus* Caltagirone (Encyrtidae, Hymenoptera)." *Journal of Experimental Zoology* 237 (2005): 309–18.

Jameson, E. W., Jr. *Patterns of Vertebrate Biology.* New York: Springer-Verlag, 1981.

Jordan, W. H., Jr. "The Weevil and the Wasp." *Natural History* 88 (October 1979): 37–43.

Quicke, D. L. J. *Parasitic Wasps.* New York: Chapman & Hall, 1997.

Reichard, M., M. Ondrackova, M. Przybylski, H. Lius, and C. Smith. "The Costs and Benefits in an Unusual Symbiosis: Experimental Evidence that Bitterling Fish (*Rhodeus sericeus*) Are Parasites of Unionid Mussels in Europe." *Journal of Evolutionary Biology* 19 (2006): 788–96.

Smith, C., M. Reichard, P. Jurajda, and M. Przybylski. "The Reproductive Ecology of the European Bitterling (*Rhodeus sericeus*)." *Journal of Zoology* 262 (2004): 107–24.

Trott, L. B. "A General Review of the Pearlfishes (Pisces, Carapidae)." *Bulletin of Marine Sciences* 31 (1981): 623–29.

Yanoviak, S. P., M. Kaspari, R. Dudley, and G. Poinar Jr. "Parasite-Induced Fruit Mimicry in a Tropical Canopy Ant." *American Naturalist* 171 (2008): 536–44.

Raising the Devil's Spawn

Baba, R., Y. Nagata, and S. Yamagishi. "Brood Parasitism and Egg Robbing among Three Freshwater Fish." *Animal Behaviour* 40 (1990): 776–78.

Brooke, M. "Tricks of the Egg Trade." *Natural History* (April 1989): 51–54.

Buschinger, A. "Evolution of Social Parasitism in Ants." *Trends in Ecology and Evolution* 1 (1986): 155–60.

Hölldobler, B., and E. O. Wilson. *The Ants.* Cambridge, MA: Belknap Press of Harvard University Press, 1990.

———. *Journey to the Ants: A Story of Scientific Exploration.* Cambridge, MA: Belknap Press of Harvard University Press, 1994.

Lembke, J. *Despicable Species: On Cowbirds, Kudzu, Hornworms, and Other Scourges.* New York: Lyons Press, 1999.

Orians, G. H. *Blackbirds of the Americas.* Seattle: University of Washington Press, 1985.

Ortega, C. P. *Cowbirds and Other Brood Parasites.* Tucson: University of Arizona Press, 1998.

Payne, R. B. "The Ecology of Brood Parasitism in Birds." *Annual Review of Ecology and Systematics* 8 (1977): 1–28.

Sato, T. "A Brood Parasitic Catfish of Mouthbrooding Cichlid Fishes in Lake Tanganyika." *Nature* 323 (1986): 58–59.

Defense Contracts

Caldwell, J. P. "The Evolution of Myrmecophagy and Its Correlates in Poison Frogs (Family Dendrobatidae)." *Journal of Zoology* (London) 240 (1996): 75–101.

Carr, A. *A Naturalist in Florida: A Celebration of Eden.* New Haven, CT: Yale University Press, 1994.

Daly, J. W., H. M. Garraffo, T. F. Spande, C. Jaramillo, and A. S. Rand. "Dietary Source for Skin Alkaloids of Poison Frogs (Dendrobatidae)?" *Journal of Chemical Ecology* 20 (1994): 943–55.

Donnelly, M. A. "Feeding Patterns of the Strawberry Poison Frog, *Dendrobates pumilio* (Anura: Dendrobatidae)." *Copeia* (1991): 723–30.

Fautin, D. G. "Sexual Stunts of Clownfish." *Natural History* (September 1989): 43–47.

———. "A Shell with a New Twist." *Natural History* (April 1992): 50–57.

Groom, M. "Sand-Colored Nighthawks Parasitize the Antipredator Behavior of Three Nesting Bird Species." *Ecology* 73 (1992): 785–93.

Haemig, P. D. "Symbiotic Nesting of Birds with Formidable Animals: A Review with Applications to Biodiversity Conservation." *Biodiversity and Conservation* 10 (2001): 527–40.

Hölldobler, B., and E. O. Wilson. *The Ants.* Cambridge, MA: Belknap Press of Harvard University Press, 1990.

———. *Journey to the Ants.* Cambridge, MA: Belknap Press of Harvard University Press, 1994.

Saporito, R. A., M. A. Donnelly, R. A. Norton, H. M. Garraffo, T. F. Spande, and J. W. Daly. "Orabatid Mites as a Major Dietary Source for Alkaloids in Poison Frogs." *Proceedings of the National Academy of Sciences* 104 (2007): 8885–90.

Simon, M. P., and C. A. Toft. "Diet Specialization in Small Vertebrates: Mite-Eating in Frogs." *Oikos* 61 (1991): 263–78.

Ueta, M. "Azure-Winged Magpies, *Cyanopica cyana*, 'Parasitize' Nest Defence Provided by Japanese Lesser Sparrowhawks, *Accipiter gularis*." *Animal Behaviour* 48 (1994): 871–74.

Wheeler, W. M. *Social Life among the Insects.* New York: Harcourt, Brace & World, 1923.

Cow Pie No. 5

Auffenberg, W. *The Behavioral Ecology of the Komodo Monitor.* Gainesville: University Presses of Florida, 1981.

Brockie, R. "Self-Anointing by Wild Hedgehogs, *Erinaceus europaeus*, in New Zealand." *Animal Behavior* 24 (1976): 68–71.

Brodie, E. D., Jr. "Hedgehogs Use Toad Venom in Their Own Defence." *Nature* 268 (1977): 627–28.

Budiansky, S. *The Truth about Dogs: An Inquiry Into the Ancestry, Social Conventions, Mental Habits, and Moral Fiber of Canis familiaris.* New York: Viking, 2000.

Clucas, B., M. P. Rowe, D. H. Owings, and P. C. Arrowood. "Snake Scent Application in Ground Squirrels, *Spermophilus* spp.: A Novel Form of Antipredator Behaviour?" *Animal Behaviour* 75 (2008): 299–307.

Kobayashi, T., and M. Watanabe. "An Analysis of Snake-Scent Application Behaviour in Siberian Chipmunks (*Eutamias sibiricus asiaticus*)." *Ethology* 72 (1986): 40–52.

Revis, H. C., and D. A. Waller. "Bactericidal and Fungicidal Activity of Ant Chemicals on Feather Parasites: An Evaluation of Anting Behavior as a Method of Self-Medication in Songbirds." *The Auk* 121 (2004): 1262–68.

Valderrama, X., J. G. Robinson, A. B. Attygalle, and T. Eisner. "Seasonal Anointment with Millipedes in a Wild Primate: A Chemical Defense against Insects?" *Journal of Chemical Ecology* 26 (2000): 2781–90.

Weldon, P. J. "Defensive Anointing: Extended Chemical Phenotype and Unorthodox Ecology." *Chemoecology* 14 (2004): 1–4.

Weldon, P. J., J. R. Aldrich, J. A. Klun, J. E. Oliver, and M. Debboun. "Benzoquinones from Millipedes Deter Mosquitoes and Elicit Self-Anointing in Capuchin Monkeys (*Cebu* spp.)." *Naturwissenschaften* 90 (2003): 301–4.

Audacious Pirates and Sneaky Burglars

Brands, H. W. *The First American: The Life and Times of Benjamin Franklin.* New York: Doubleday, 2000.

Burger, A. E. "Time Budgets, Energy Needs and Kleptoparasitism in Breeding Lesser Sheathbills (*Chionis minor*)." *Condor* 83 (1981): 106–12.

Eberhard, W. G. "Spider and Fly Play Cat and Mouse." *Natural History* 89 (January 1980): 56–60.

Elphick, C., J. B. Dunning Jr., and D. A. Sibley, eds. *The Sibley Guide to Bird Life and Behavior.* New York: Alfred A. Knopf, 2001.

Furness, R. W. "Kleptoparasitism in Seabirds." In *Seabirds: Feeding Ecology and Role in Marine Ecosystems,* ed. J. P. Croxall, pp. 77–100. Cambridge: Cambridge University Press, 1987.

Hatch, J. J. "Predation and Piracy by Gulls at a Ternery in Maine." *The Auk* 87 (1970): 244–54.

Hölldobler, B., and E. O. Wilson. *The Ants.* Cambridge, MA: Belknap Press of Harvard University Press, 1990.

Nelson, B. *Galapagos: Islands of Birds.* New York: William Morrow, 1968.

Nelson, J. B. "The Breeding Biology of Frigatebirds—A Comparative Review." *Living Bird* 14 (1975): 113–56.

Oldroyd, H. *The Natural History of Flies.* New York: W. W. Norton, 1964.

O'Toole, C., ed. *Firefly Encyclopedia of Insects and Spiders.* Buffalo, NY: Firefly Books, 2002.

Vander Wall, S. B. *Food Hoarding in Animals.* Chicago: University of Chicago Press, 1990.

Wilson, E. O. *Sociobiology: The New Synthesis.* Cambridge, MA: Belknap Press of Harvard University Press, 1975.

Sexy Orchids Make Lousy Lovers, and Other Orchid Contrivances

Bembé, B. "Functional Morphology in Male Euglossine Bees and Their Ability to Spray Fragrances (Hymenoptera, Apidae, Euglossini)." *Apidologie* 35 (2004): 283–91.

Dafni, A. "Mimicry and Deception in Pollination." *Annual Review of Ecology and Systematics* 15 (1984): 259–78.

Darwin, C. *The Various Contrivances by Which Orchids Are Fertilized by Insects.* 2nd ed., rev. Chicago: University of Chicago Press, 1984. Originally published in 1877.

Dressler, R. L. "Biology of the Orchid Bees (Euglossini)." *Annual Review of Ecology and Systematics* 13 (1982): 373–94.

Hansen, E. "Bee Bop." *Natural History* 108 (March 1999): 72–74.

Jersáková, J., S. D. Johnson, and P. Kindlmann. "Mechanisms and Evolution of Deceptive Pollination in Orchids." *Biological Reviews* 81 (2006): 219–35.

Van der Pijl, L., and C. H. Dodson. *Orchid Flowers: Their Pollination and Evolution.* Coral Gables, FL: Fairchild Tropical Garden and the University of Miami Press, 1966.

A Seedy Neighborhood

Beattie, A. J. "Ant Plantation." *Natural History* 99 (February 1990): 10–14.

Dunn, R. R. "Jaws of Life." *Natural History* 114 (September 2005): 30–35.

Fialho, R. "Seed Dispersal by a Lizard and a Treefrog—Effect of Dispersal Site on Seed Survivorship." *Biotropica* 22 (1990): 423–24.

Gottsberger, G. "Seed Dispersal by Fish in the Inundated Regions of Humaitá, Amazonia." *Biotropica* 10 (1978): 170–83.

Iverson, J. B. "Lizards as Seed Dispersers?" *Journal of Herpetology* 19 (1985): 292–93.

Kubitzki, K., and A. Ziburski. "Seed Dispersal in Flood Plain Forests of Amazonia." *Biotropica* 26 (1994): 30–43.

Liu, H., S. G. Platt, and C. K. Borg. "Seed Dispersal by the Florida Box Turtle (*Terrapene carolina bauri*) in Pine Rockland Forests of the Lower Florida Keys, United States." *Oecologia* 138 (2004): 539–46.

Silva, H. R. da, M. C. de Britto-Pereira, and U. Caramaschi. "Frugivory and Seed Dispersal by *Hyla truncata*, a Neotropical Treefrog." *Copeia* (1989): 781–83.

Thoreau, H. D. *Faith in a Seed: The Dispersion of Seeds and Other Late Natural History Writings.* Washington, DC: Island Press/Shearwater Books, 1993.

Vander Wall, S., and R. P. Balda. "Remembrance of Seeds Stashed." *Natural History* 92 (September 1983): 61–65.

Green, Green Plants of Home

Bradshaw, W., and C. Holzapfel. "Life in a Deathtrap." *Natural History* 100 (July 1991): 35–36.

Crump, M. L. "Parental Care." In *Amphibian Biology.* Vol. 2, *Social Behaviour,* ed. H. Heatwole and B. K. Sullivan, pp. 518–67. Chipping Norton, NSW: Surrey Beatty & Sons, 1995.

Diesel, R. "Managing the Offspring Environment: Brood Care in the Bromeliad Crab, *Metopaulias depressus.*" *Behavioral Ecology and Sociobiology* 30 (1992): 125–34.

Diesel, R., and M. Schuh. "Maternal Care in the Bromeliad Crab *Metopaulias depressus* (Decapoda): Maintaining Oxygen, pH and Calcium Levels Optimal for the Larvae." *Behavioral Ecology and Sociobiology* 32 (1993): 11–15.

Hölldobler, B., and E. O. Wilson. *The Ants.* Cambridge, MA: Belknap Press of Harvard University Press, 1990.

Janzen, D. H. "*Blastophaga* and Other Agaonidae." In *Costa Rican Natural History,* ed. D. H. Janzen, pp. 696–700. Chicago: University of Chicago Press, 1983.

———. "How to Be a Fig." *Annual Review of Ecology and Systematics* 10 (1979): 13–51.

Jones, R. W. "March of the Weevils." *Natural History* 115 (February 2006): 30–35.

Moran, J. "Life and Death in a Pitcher." *Natural History* 115 (October 2006): 56–62.

Timm, R. M., and B. L. Clauson. "A Roof Over Their Feet." *Natural History* 99 (March 1990): 55–59.

Wayne's Word Noteworthy Plant for August 1997. Mexican Jumping Beans. http://waynesword.palomar.edu/plaug97.htm.

Weygoldt, P. "Complex Brood Care and Reproductive Behavior in Captive Poison-Arrow Frogs, *Dendrobates pumilio* O. Schmidt." *Behavioral Ecology and Sociobiology* 7 (1980): 329–32.

Powerful Plant Products

Crump, M. *In Search of the Golden Frog.* Chicago: University of Chicago Press, 2000.

Davis, W. *One River: Explorations and Discoveries in the Amazon Rain Forest.* New York: Touchstone, 1996.

Dupain, J., L. Van Elsacker, C. Nell, P. Garcia, F. Ponce, and M. A. Huffman. "New Evidence for Leaf Swallowing and *Oesophagostomum* Infection in Bonobos (*Pan paniscus*)." *International Journal of Primatology* 23 (2002): 1053–62.

Engel, C. *Wild Health: How Animals Keep Themselves Well and What We Can Learn from Them.* Boston: Houghton Mifflin, 2002.

Mann, J. *Murder, Magic and Medicine.* New York: Oxford University Press, 2000.

Plotkin, M. J. *Tales of a Shaman's Apprentice.* New York: Viking, 1993.

Slotkin, J. *The Peyote Religion.* Glencoe, IL: Free Press, 1956.

There's the Rub

Andrews, R. "Western Science Learns from Native Culture." *The Scientist* (March 16, 1992). http://www.the-scientist.com/yr1992/march/andrews_p6_920316.html.

Baker, M. "Fur Rubbing: Use of Medicinal Plants by Capuchin Monkeys (*Cebus capucinus*)." *American Journal of Primatology* 38 (1996): 263–70.

Clayton, D. H., and J. G. Vernon. "Common Grackle Anting with Lime Fruit and Its Effect on Ectoparasites." *The Auk* 110 (1993): 951–52.

Engel, C. "Bear Self-Medication." *International Bear News* 11 (February 2002): 34–35.

Laska, M., V. Bauer, and L. T. Hernandez Salazar. "Self-Anointing Behavior in Free Ranging Spider Monkeys (*Ateles geoffroyi*) in Mexico." *Primates* 48 (2007): 160–63.

Ants and Plants

Beattie, A. J. *The Evolutionary Ecology of Ant-Plant Mutualisms.* Cambridge: Cambridge University Press, 1985.

———. "Myrmecotrophy: Plants Fed by Ants." *Trends in Ecology and Evolution* 4 (1989): 172–76.

Belt, T. *The Naturalist in Nicaragua.* Chicago: University of Chicago Press, 1985. Originally published, London: J. Murray, 1874.

Hölldobler, B., and E. O. Wilson. *The Ants.* Cambridge, MA: Belknap Press of Harvard University Press, 1990.

Huxley, C. R., and D. F. Cutler, eds. *Ant-Plant Interactions.* Oxford: Oxford University Press, 1991.

Janzen, D. H. "Epiphytic Myrmecophytes in Sarawak: Mutualism through the Feeding of Plants by Ants." *Biotropica* 6 (1974): 237–52.

Korndorfer, A. P., and K. Del-Claro. "Ant Defense versus Induced Defense in *Lafoensia pacari* (Lythraceae), a Myrmecophilous Tree of the Brazilian Cerrado." *Biotropica* 38 (2006): 786–88.

Rico-Gray, V., and P. S. Oliveira. *The Ecology and Evolution of Ant-Plant Interactions.* Chicago: University of Chicago Press, 2007.

Rico-Gray, V., J. T. Barber, L. B. Thien, E. G. Ellgaard, and J. J. Toney. "An Unusual Animal-Plant Interaction: Feeding of *Schomburgkia tibicinis* (Orchidaceae) by Ants." *American Journal of Botany* 76 (1989): 603–8.

Solano, P. J., and A. Dejean. "Ant-Fed Plants: Comparison between Three Geophytic Myrmecophytes." *Biological Journal of the Linnean Society* 83 (2004): 433–39.

Thompson, J. N. "Reversed Animal-Plant Interactions: The Evolution of Insectivorous and Ant-Fed Plants." *Biological Journal of the Linnean Society* 16 (1981): 147–55.

Tilman, D. "Cherries, Ants and Tent Caterpillars: Timing of Nectar Production in Relation to Susceptibility of Caterpillars to Ant Predation." *Ecology* 59 (1978): 686–92.

Treseder, K. K., D. W. Davidson, and J. R. Ehleringer. "Absorption of Ant-Provided Carbon Dioxide and Nitrogen by a Tropical Epiphyte." *Nature* 375 (1995): 137–39.

Intestinal Microbes and the Gas We Pass

Grajal, A., and S. D. Strahl. "A Bird with the Guts to Eat Leaves." *Natural History* (August 1991): 48–55.

Grajal, A., S. D. Strahl, R. Parra, M. G. Domínguez, and A. Neher. "Foregut Fermentation in the Hoatzin, a Neotropical Leaf-Eating Bird." *Science* 245 (1989): 1236–38.

Lee, A., and R. Martin. "Life in the Slow Lane." *Natural History* (August 1990): 34–43.

Postgate, J. *Microbes and Man.* 4th ed. Cambridge: Cambridge University Press, 2000.

Pough, F. H., C. M. Janis, and J. B. Heiser. *Vertebrate Life.* 6th ed. Upper Saddle River, NJ: Prentice Hall, 2002.

Troyer, K. "Transfer of Fermentative Microbes between Generations in a Herbivorous Lizard." *Science* 216 (1982): 540–42.

Wakeford, T. *Liaisons of Life: From Hornworts to Hippos, How the Unassuming Microbe Has Driven Evolution.* New York: John Wiley & Sons, 2001.

Deadly Dragon Drool

Auffenberg, W. *The Behavioral Ecology of the Komodo Monitor.* Gainesville: University Presses of Florida, 1981.

———. "Komodo Dragons." *Natural History* 81 (April 1972): 52–59.

Burden, W. D. "Stalking the Dragon Lizard on the Island of Komodo." *National Geographic* 52 (February 1927): 216–33.

Cheater, M. "Chasing the Magic Dragon." *National Wildlife Magazine* 41 (August/September 2003).

Gillespie, D., T. Fredeking, and J. M. Montgomery. "Microbial Biology and Immunology." In *Komodo Dragons: Biology and Conservation,* ed. J. B. Murphy, C. Ciofi, C. de La Panouse, and T. Walsh, pp. 118–26. Washington, DC: Smithsonian Institution Press, 2002.

Kern, J. A. "Dragon Lizards of Komodo." *National Geographic* (December 1968): 874–80.

Montgomery, J. M., D. Gillespie, P. Sastrawan, T. M. Fredeking, and G. L. Stewart. "Aerobic Salivary Bacteria in Wild and Captive Komodo Dragons." *Journal of Wildlife Diseases* 38 (2002): 545–51.

Murphy, J. B., C. Ciofi, C. de La Panouse, and T. Walsh, eds. *Komodo Dragons: Biology and Conservation.* Washington, DC: Smithsonian Institution Press, 2002.

Price, M. E. *Myths and Enchantment Tales.* New York: Rand McNally, 1953.

Shnayerson, M., and M. J. Plotkin. *The Killers Within: The Deadly Rise of Drug-Resistant Bacteria.* Boston: Little, Brown, 2002.

Sweet, S. S., and E. R. Pianka. "The Lizard Kings." *Natural History* 112 (November 2003): 40–45.

Mighty Mushrooms and Other Good Fungus among Us

Barchfield, J. "Truffle Trouble." *Arizona Daily Sun* (February 24, 2008), p. A8.

Damrosch, B. "Corn Smut: A Reputation Redeemed." *Washington Post* (February 15, 2007), p. H8.

Erikson, P. "Near Beer of the Amazon." *Natural History* (August 1990): 52–61.

Ingram, C. *Vegetarian and Vegetable Cooking.* New York: Hermes House, 2002.

Moore, D. *Slayers, Saviors, Servants, and Sex: An Exposé of Kingdom Fungi.* New York: Springer-Verlag, 2001.

Postgate, J. *Microbes and Man.* 4th ed. Cambridge: Cambridge University Press, 2000.

Selosse, M.-A., E. Baudoin, and P. Vandenkoornhuyse. "Symbiotic Microorganisms: A Key for Ecological Success and Protection of Plants." *C. R. Biologies* 327 (2004): 639–48.

Wakeford, T. *Liaisons of Life: From Hornworts to Hippos, How the Unassuming Microbe Has Driven Evolution.* New York: John Wiley & Sons, 2001.

Wang, B., and Y.-L. Qiu. "Phylogenetic Distribution and Evolution of Mycorrhizas in Land Plants." *Mycorrhiza* 16 (2006): 299–363.

Bombarded by Bacteria

Bellis, M. "The History of Penicillin: Alexander Fleming, John Sheehan, Andrew J. Moyer." http://inventors.about.com/od/pstartinventions/a/Penicillin.htm.

Burdick, A. E., and I. D. Camacho. "Impetigo." July 15, 2008. www.emedicine.com/derm/topic195.htm.

Cohen, J. O., ed. *The Staphylococci.* New York: Wiley-Interscience, 1972.

"Conjunctivitis (Pink Eye)." St. Luke's Cataract and Laser Institute. www.stlukeseye.com/Conditions/Conjunctivitis.asp.

Freeman, S. "Bacterial Diseases." In *Biological Science,* pp. 495–98. Upper Saddle River, NJ: Prentice Hall, 2002.

Hegner, R. W. "Adam Had 'Em." In *Nature Smiles in Verse,* ed. B. R. Weimer. Baltimore, MD: Waverly Press, 1940.

"Impetigo." DermNet NZ. www.dermnetnz.org/bacterial/impetigo.html.

"Impetigo." Health A to Z. www.healthatoz.com/healthatoz/Atoz/ency/impetigo.jsp.

Lewis, R. "The Rise of Antibiotic Resistant Infections." *FDA Consumer,* September 1995. www.fda.gov/fdac/features/795_antibio.html.

Madigan, M. T., and J. M. Martinko. *Biology of Microorganisms.* 11th ed. Upper Saddle River, NJ: Prentice Hall, 2006.

Marlin, D. S. "Conjunctivitis, Bacterial." May 10, 2007. www.emedicine.com/OPH/topic88.htm.

Patrick, S., and M. J. Larkin. *Immunological and Molecular Aspects of Bacterial Virulence.* New York: John Wiley & Sons, 1995.

Singleton, P. *Bacteria in Biology, Biotechnology and Medicine.* 3rd ed. New York: John Wiley & Sons, 1995.

"Sinus Infections (Sinusitis)." National Institute of Allergy and Infectious Diseases, National Institutes of Health. www.niaid.nih.gov/factsheets/sinusitis.htm.

Skinner, F. A., and L. B. Quesnel, eds. *Streptococci.* New York: Academic Press, 1978.

A Cloak of Antibiotics

Austin, R. M. "Cutaneous Microbial Flora and Antibiosis in *Plethodon ventralis.*" In *The Biology of Plethodontid Salamanders,* ed. R. C. Bruce, R. G. Jaeger, and L. D. Houck, pp. 127–36. New York: Kluwer Academic/Plenum, 2000.

Currie, C. R., J. A. Scott, R. C. Summerbell, and D. Malloch. "Fungus-Growing Ants Use Antibiotic-Producing Bacteria to Control Garden Parasites." *Nature* 398 (1999): 701–4.

Currie, C. R., B. Wong, A. E. Stuart, T. R. Schultz, S. A. Rehner, U. G. Mueller, G.-H. Sung, J. W. Spatafora, and N. A. Straus. "Ancient Tripartite Coevolution in the Attine Ant-Microbe Symbiosis." *Science* 299 (2003): 386–88.

Harris, R. N., T. Y. James, A. Lauer, M. A. Simon, and A. Patel. "Amphibian Pathogen *Batrachochytrium dendrobatidis* Is Inhibited by the Cutaneous Bacteria of Amphibian Species." *EcoHealth* 3 (2006): 53–56.

Hölldobler, B., and E. O. Wilson. *The Ants.* Cambridge, MA: Belknap Press of Harvard University Press, 1990.

Kaltenpoth, M., W. Gottler, G. Herzner, and E. Strohm. "Symbiotic Bacteria Protect Wasp Larvae from Fungal Infestation." *Current Biology* 15 (2005): 475–79.

Poulsen, M., A. N. M. Bot, C. R. Currie, M. G. Nielsen, and J. J. Boomsma. "Within-Colony Transmission and the Cost of a Mutualistic Bacterium in the Leaf-Cutting Ant *Acromyrmex octospinosus.*" *Functional Ecology* 17 (2003): 260–69.

Silliman, B. R., and S. Y. Newell. "Fungal Farming in a Snail." *Proceedings of the National Academy of Sciences* 100 (2003): 15643–48.

Stevens, G. C. "*Atta cephalotes* (Zompopas, Leaf-Cutting Ants)." In *Costa Rican Natural History,* ed. D. H. Janzen, pp. 688–91. Chicago: University of Chicago Press, 1983.

Strohm, E., and E. Linsenmair. "Females of the European Beewolf Preserve Their Honeybee Prey against Competing Fungi." *Ecological Entomology* 26 (2001): 198–203.

Invasion of the Body Snatchers

Brodie, H. J. *Fungi: Delight of Curiosity.* Toronto: University of Toronto Press, 1978.

Crump, M. *In Search of the Golden Frog.* Chicago: University of Chicago Press, 2000.

Fungi Perfecti. www.fungiperfecti.com.

Holliday, J., and M. Cleaver. "On the Trail of the Yak: Ancient Cordyceps in the Modern World." June 2004.http://pharmaceuticalmushrooms.nwbotanicals.org/lexicon/cordyceps/cordy_chapter%20IJM.pdf.

Ingold, C. T., and H. J. Hudson. *The Biology of Fungi*. 6th ed. New York: Chapman & Hall, 1993.

Samson, R. A., H. C. Evans, and J.-P. Latgé. *Atlas of Entomopathogenic Fungi*. New York: Springer-Verlag, 1988.

Shah, P. A., and J. K. Pell. "Entomopathogenic Fungi as Biological Control Agents." *Applied Microbiology and Biotechnology* 61 (2003): 413–23.

Steinkraus, D. C., and J. B. Whitfield. "Chinese Caterpillar Fungus and World Record Runners." *American Entomologist* 40 (1994): 235–39.

Body Snatchers Revisited

Angus, S., trans. *The Mystery Religions and Christianity* (London, 1925), p. 140. (Quote of Scholiast on Aristophanes.)

Baskett, T. F. "A Flux of the Reds: Evolution of Active Management of the Third Stage of Labour." *Journal of the Royal Society of Medicine* 93 (2000): 489–93.

Caporael, L. R. "Ergotism: The Satan Loosed in Salem?" *Science* 192 (1976): 21–26.

Ingold, C. T., and H. J. Hudson. *The Biology of Fungi*. 6th ed. New York: Chapman & Hall, 1993.

Mann, J. *Murder, Magic and Medicine*. New York: Oxford University Press, 2000.

Moore, D. *Slayers, Saviors, Servants, and Sex: An Exposé of Kingdom Fungi*. New York: Springer-Verlag, 2001.

Schiff, P. L. "Ergot and Its Alkaloids." *American Journal of Pharmacology Education* 70 (2006): 1–10.

Valencic, I. "Has the Mystery of the Eleusinian Mysteries Been Solved?" *Yearbook for Ethnomedicine and the Study of Consciousness* 3 (1994): 325–36.

Wasson, R. G., A. Hofmann, and C. A. P. Ruck. *The Road to Eleusis*. New York: Harcourt Brace Jovanovich, 1978.

Conserve Interactions, Not Just Species

Bailey, J. K., and T. G. Whitham. "Interactions between Cottonwood and Beavers Positively Affect Sawfly Abundance." *Ecological Entomology* 31 (2006): 294–97.

Bailey, J. K., J. A. Schweitzer, B. J. Rehill, R. L. Lindroth, G. D. Martinsen, and T. G. Whitham. "Beavers as Molecular Geneticists: A Genetic Basis to the Foraging of an Ecosystem Engineer." *Ecology* 85 (2004): 603–8.

Janzen, D, H. "Promising Directions of Study in Tropical Animal-Plant Interactions." *Annals of the Missouri Botanical Garden* 64 (1977): 706–36.

Martinsen, G. D., E. M. Driebe, and T. G. Whitham. "Indirect Interactions Mediated by Changing Plant Chemistry: Beaver Browsing Benefits Beetles." *Ecology* 79 (1998): 192–200.

Noss, R. F., and A. Y. Cooperrider. *Saving Nature's Legacy: Protecting and Restoring Biodiversity*. Washington, DC: Island Press, 1994.

Index